Inequality by Design

✜

Inequality by Design

CRACKING THE BELL CURVE MYTH

CLAUDE S. FISCHER, MICHAEL HOUT,

MARTÍN SÁNCHEZ JANKOWSKI,

SAMUEL R. LUCAS,

ANN SWIDLER,

AND KIM VOSS

Department of Sociology
University of California, Berkeley

PRINCETON UNIVERSITY PRESS

PRINCETON, NEW JERSEY

Library of Congress Cataloging-in-Publication Data
Inequality by design : cracking the
bell curve myth / Claude S. Fischer . . . [et al.].
p.   cm.
Includes bibliographical references and index.
ISBN 0-691-02899-0 (cl : alk. paper). —
ISBN 0-691-02898-2 (pb : alk. paper)
1. Intellect.   2. Nature and nurture.
3. Intelligence levels—United States.
4. Intelligence levels—Social aspects—
United States.   5. Educational psychology.
6. Herrnstein, Richard J. Bell curve.
I. Fischer, Claude S., 1948–
BF431.I513   1996
305.9′082—dc20   96-2171   CIP

This book has been composed in Times Roman

Princeton University Press books are printed
on acid-free paper and meet the guidelines
for permanence and durability of the Committee
on Production Guidelines for Book Longevity
of the Council on Library Resources

http://pup.princeton.edu

Printed in the United States of America

3 5 7 9 10 8 6 4

5 7 9 10 8 6
(Pbk.)

TO OUR CHILDREN

✠

We have been quick to seek explanations of our problems and failures in what we *are* instead of what we *do*. We seem wedded to the belief that our situation is a consequence of our nature rather than of our historical acts . . .

—Kenneth Bock, *Human Nature Mythology*

❖ *Contents* ❖

❖ *Figures and Tables* ❖

FIGURES

TABLES

✤ Preface ✤

WE WERE impelled to write this book by the publication in late 1994 of *The Bell Curve*. That immensely well publicized book was then the latest statement of a philosophy that gained extensive credence in the 1990s: The widening inequalities among Americans that developed in the last quarter-century are inevitable. Because of human nature, because of the nature of the market, because of the nature of modern society, Americans will necessarily divide more and more by social class and race. We reject that philosophy. Besides being morally complacent, it is a doctrine without scientific foundation. Research has shown that "nature" determines neither the level of inequality in America nor which Americans in particular will be privileged or disprivileged; social conditions and national policies do. Inequality is in that sense designed. Similarly, the market does not require us to accept great inequalities for the sake of growth. Quite the reverse seems true; nations do better the more equal their citizens are. And modern societies can be less unequal than America is in 1996; others are now and ours has been in the past. *The Bell Curve*, in particular, as an emphatic statement of this mistaken philosophy, is wrong to claim that differences in native intelligence explain inequality. The social science evidence is clear. We see our task as bringing such evidence to the attention of the wider public.

In late 1994, we—all members of Berkeley's Department of Sociology—came together to discuss *The Bell Curve* phenomenon and soon agreed that a response from sociologists was in order. Some colleagues suggested that *The Bell Curve* would fade from public consciousness. After all, its arguments about the significance of intelligence had already been dealt with. (Almost a quarter-century ago, Christopher Jencks and his colleagues showed, in *Inequality*, that individuals' intelligence at best only modestly affects their fortunes.) But the ideology *The Bell Curve* represents is too pervasive; the book's shock waves are too great to ignore. As social scientists, we feel responsible for correcting the record. As university teachers, we are painfully aware that *The Bell Curve* has unsettled our students. As citizens, we must participate in the national debate. So we set aside much of our ongoing work to write this book.

Inequality by Design is a true collaboration. While particular individuals took the lead in drafting specific chapters, everyone joined in outlining the basic argument, contributing ideas, and revising drafts. Fischer had

xi

responsibility, in addition, for coordinating the project and giving the book a single authorial voice.

We benefited from the collaboration as well of several Berkeley graduate students. Richard Arum was so critical to our reanalysis of *The Bell Curve* data that he shares authorship of chapters 3 and 4. Elizabeth Armstrong, Leslie Bell, Charlotte Chiu, Tally Katz, Amy Schalet, and Sean Stryker helped us find some of the scientific literature. Berkeley staff who contributed include Judy Haier and Maureen Fesler. We also had the "collaboration" of several scholars who gave us either suggestions or comments, or both. We thank Robert Bellah, Fred Block, Joe Harder, Adam Hochschild, Christopher Jencks, Rob Macoun, Douglas Massey, Lee Rainwater, Paul Romer, Saul Rubin, David Vogel, and Alan Wolfe. We especially thank David Levine for commenting closely on an early draft. William Dickens, Troy Duster, Robert Hauser, and Chris Winship shared their prepublication manuscripts with us. Audiences who heard early versions of the work at the University of Arizona, University of Virginia, Princeton University, the University of California, Berkeley, and the May 1995 meeting of the International Sociological Association's Research Committee on Stratification and Mobility helped hone our arguments.

No large grants funded this work, but Berkeley's Department of Sociology and Survey Research Center provided meeting rooms and occasional secretarial assistance. In addition, the Survey Research Center and the Institute of Industrial Relations supported a few of the graduate students who helped us.

We also very much appreciate the commitment and energy that Peter Dougherty, publisher of Social Science and Public Affairs at Princeton University Press, gave this book. We also thank, at the Press, Michelle McKenna and Jane Low. Anita O'Brien was our amazingly efficient copy editor. Chris Brest drew the figures.

We hope that this book will help redirect the public discussion away from the mistaken, helpless view that inequality is fated or necessary and toward the more accurate, empowering understanding that opportunity and equality are very much within citizens' control. We look forward to a nation that, as great as it is today, will be a yet fairer one when our children, to whom we dedicate this work, shoulder the burdens of its citizenship.

Inequality by Design

✣

Why Inequality?

As we write, Americans are engaged in a great debate about the inequalities that increasingly divide us. For over twenty years, the economic gaps have widened. As the American Catholic Bishops stated in late 1995, "the U.S. economy sometimes seems to be leading to three nations living side by side, one growing more prosperous and powerful, one squeezed by stagnant incomes and rising economic pressures and one left behind in increasing poverty, dependency and hopelessness."[1] Being prosperous may mean owning a vacation home, purchasing private security services, and having whatever medical care one wants; being squeezed may mean having one modest but heavily mortgaged house, depending on 911 when danger lurks, and delaying medical care because of the expense of copayments; and being left behind may mean barely scraping together each month's rent, relying on oneself for physical safety, and awaiting emergency aid at an overcrowded public clinic. Most Americans in the middle know how fragile their position is. One missed mortgage payment or one chronic injury might be enough to push them into the class that has been left behind.

Few deny that inequality has widened.[2] The debate is over whether anything can be done about it, over whether anything should be done about it. Some voices call for an activist government to sustain the middle class and uplift the poor. Other voices, the ones that hold sway as we write, argue that government ought to do less, not more. They argue for balanced budgets, lower taxes, fewer domestic programs, minimal welfare, and less regulation. These moves, they contend, would energize the economy and in that way help the middle class. They would also help the poor, economically and otherwise. Speaker of the House Newt Gingrich in 1995 said of people on welfare: "The government took away something more important than . . . money. They took away their initiative, . . . their freedom, . . . their morality, their drive, their pride. I want to help them get that back."[3] As to the increasing inequality of our time, some advocates of circumscribed government say it cannot be changed, because inequality is natural; some say it ought not be changed, because inequality drives our economy. At a deeper level, then, the debate is about how to understand inequality—what explains its origin, what explains its growth. That is where we shall engage the debate.

The arguments over policy emerged from almost a quarter century of economic turmoil and disappointment. Middle-class Americans saw the era of seemingly ever-expanding affluence for themselves and ever-expanding opportunities for their children come to an abrupt end in 1973. The cars inching forward in the gasoline lines of the mid-1970s foreshadowed the next twenty years of middle-class experience. Wages stagnated, prices rose, husbands worked longer hours, and even wives who preferred to stay at home felt pressed to find jobs. The horizons for their children seemed to shrink as the opportunities for upward economic mobility contracted.[4] What was going on? What could be done about it?

In the early 1980s, one explanation dominated public discussion and public policy: The cause of the middle-class crisis was government, and its solution was less government. Regulations, taxes, programs for the poor, preferences for minorities, spending on schools—indeed, the very size of government—had wrecked the economy by wasting money and stunting initiative, by rewarding the sluggards and penalizing the talented. The answer was to get government "off the backs" of those who generate economic growth. "Unleash the market" and the result would be a "rising tide that will lift all boats, yachts and rowboats alike."

This explanation for the economic doldrums won enough public support to be enacted. Less regulation, less domestic spending, and more tax cuts for the wealthy followed. By the 1990s, however, the crisis of the middle class had not eased; it had just become more complicated. Figure 1.1 shows the trends in family incomes, adjusted for changes in prices, from 1959 to 1989 (the trends continued into the 1990s). The richest families had soared to new heights of income, the poorest families had sunk after 1970, and the middle-income families had gained slightly. But this slight gain was bitterly misleading. The middle class managed to sustain modest income growth only by mothers taking jobs and fathers working longer hours. Also, the slight gain could not make up for growing economic insecurity and parents' anxiety that key elements of the "American Dream"—college education, a stable job, and an affordable home—were slipping beyond the grasp of their children. And so the phrase "the disappearing middle class" began to be heard.

Another puzzle now called for explanation: The 1980s had been a boom decade; overall wealth had grown. But average Americans were working harder to stay even. Why had the gaps between the rich and the middle and between the middle and the poor widened? How do we understand such inequality?

Between 1959 and 1969, income per person grew for all households. Since 1970, income per person has continued to grow rapidly for the richest households, grown at a declining rate among middle–income households, and fallen slightly among poor house-holds. The result is significantly more inequality.

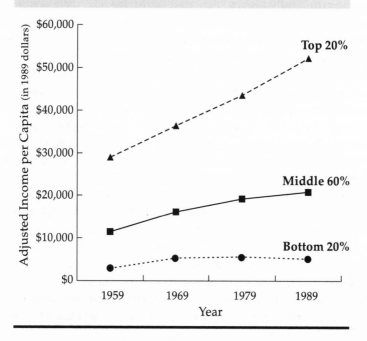

1.1. Changes in Household Incomes, 1959–1989, by Income Class (*Note*: Household incomes are adjusted by dividing income per family member by the square root of the household size. *Source*: Karoly and Burtless, "Demographic Change, Rising Earnings Inequality," table 2)

An answer emerged in the public debate, forwarded for the most part by the same voices that had offered the earlier explanation: Inequality is a "natural," almost inevitable, result of an unfettered market. It is the necessary by-product of unleashing talent. The skilled soar and the unskilled sink. Eventually, however, all will gain from the greater efficiency of the free market. The reason such wider benefits have yet to be delivered is that the market has not been freed up enough; we need still less government and then the wealth will flow to middle- and lower-income Americans. Sharp inequality among the classes, these voices suggested, is the necessary trade-off for economic growth.

The strongest recent statement that inequality is the natural result of a free market came in *The Bell Curve: Intelligence and Class Structure in American Life*, published in 1994. Richard Herrnstein and Charles Murray argued that intelligence largely determined how well people did in life. The rich were rich mostly because they were smart, the poor were poor mostly because they were dumb, and middle Americans were middling mostly because they were of middling intelligence. This had long been so but was becoming even more so as new and inescapable economic forces such as global trade and technological development made intelligence more important than ever before. In a more open economy, people rose or sank to the levels largely fixed by their intelligence. Moreover, because intelligence is essentially innate, this expanding inequality cannot be stopped. It might be slowed by government meddling, but only by also doing injustice to the talented and damaging the national economy. Inequality is in these ways "natural," inevitable, and probably desirable.

The Bell Curve also provided an explanation for another troubling aspect of inequality in America—its strong connection to race and ethnicity. Black families, for example, are half as likely to be wealthy and twice as likely to be poor as white families. The questions of how to understand racial disparities and what to do about them have anguished the nation for decades. Now, there was a new answer (actually, a very old answer renewed): Blacks—and Latinos, too—were by nature not as intelligent as whites; that is why they did less well economically, and that is why little can or should be done about racial inequality.

Yet decades of social science research, and further research we will present here, refute the claim that inequality is natural and increasing inequality is fated. Individual intelligence does not satisfactorily explain who ends up in which class; nor does it explain why people in different classes have such disparate standards of living. Instead, what better explains inequality is this: First, individuals' social milieux—family, neighborhood, school, community—provide or withhold the means for attaining higher class positions in American society, in part by providing people with marketable skills. Much of what those milieux have to offer is, in turn, shaped by social policy. For example, the quality of health care that families provide and the quality of education that schools impart are strongly affected by government action. Second, social policy significantly influences the rewards individuals receive for having attained their positions in society.

6

Circumstances—such as how much money professional or manual workers earn, how much tax they pay, whether their child care or housing is subsidized—determine professionals' versus manual workers' standards of living. In turn, these circumstances are completely or partly determined by government. We do *not* have to suffer such inequalities to sustain or expand our national standard of living.[5] Thus, inequality is not the natural and inevitable consequence of intelligence operating in a free market; in substantial measure it is and will always be the socially constructed and changeable consequence of Americans' political choices.

Our contribution to the debate over growing inequality is to clarify how and why inequality arises and persists. We initiate our argument by first challenging the explanation in *The Bell Curve*, the idea that inequality is natural and fated. Then, we go on to show how social environment and conscious policy mold inequality in America.

If the growing inequality in America is not the inevitable result of free markets operating on natural intelligence, but the aftermath of circumstances that can be altered, then different policy implications follow from those outlined in *The Bell Curve*. We do not have to fatalistically let inequalities mount; we do not have to accept them as the Faustian trade for growth; and we do not have to accept heartlessness as the companion of social analysis. Instead, we can anticipate greater equality of opportunity and equality of outcome and also greater economic growth.

Explaining Inequality

Why do some Americans have a lot more than others? Perhaps, inequality follows inevitably from human nature. Some people are born with more talent than others; the first succeed while the others fail in life's competition. Many people accept this explanation, but it will not suffice. Inequality is not fated by nature, nor even by the "invisible hand" of the market; it is a social construction, a result of our historical acts. *Americans have created the extent and type of inequality we have, and Americans maintain it.*

To answer the question of what explains inequality in America, we must divide it in two. First, who gets ahead and who falls behind in the competition for success? Second, what determines how much people get for being ahead or behind? To see more clearly that the two questions are different, think of a ladder that represents the ranking of affluence in a society. Question one asks why this person rather than that person ended up on a higher or lower rung. Question two asks why some societies have tall and narrow-

7

ing ladders—ladders that have huge distances between top and bottom rungs and that taper off at the top so that there is room for only a few people—while other societies have short and broad ladders—ladders with little distance between top and bottom and with lots of room for many people all the way to the top.

(Another metaphor is the footrace: One question is who wins and who loses; another question is what are the rules and rewards of the race. Some races are winner-take-all; some award prizes to only the first few finishers; others award prizes to many finishers, even to all participants. To understand the race, we need to understand the rules and rewards.)

The answer to the question of who ends up where is that people's social environments largely influence what rung of the ladder they end up on.[6] The advantages and disadvantages that people inherit from their parents, the resources that their friends can share with them, the quantity and quality of their schooling, and even the historical era into which they are born boost some up and hold others down. The children of professors, our own children, have substantial head starts over children of, say, factory workers. Young men who graduated from high school in the booming 1950s had greater opportunities than the ones who graduated during the Depression. Context matters tremendously.

The answer to the question of why societies vary in their structure of rewards is more political. In significant measure, societies choose the height and breadth of their "ladders." By loosening markets or regulating them, by providing services to all citizens or rationing them according to income, by subsidizing some groups more than others, societies, through their politics, build their ladders. To be sure, historical and external constraints deny full freedom of action, but a substantial freedom of action remains (see, especially, chapters 5 and 6). In a democracy, this means that the inequality Americans have is, in significant measure, the historical result of policy choices Americans—or, at least, Americans' representatives—have made. In the United States, the result is a society that is distinctively *un*equal. Our ladder is, by the standards of affluent democracies and even by the standards of recent American history, unusually extended and narrow—and becoming more so.

To see how policies shape the structure of rewards (i.e., the equality of outcomes), consider these examples: Laws provide the ground rules for the marketplace—rules covering incorporation, patents, wages, working conditions, unionization, security transactions, taxes, and so on. Some laws widen differences in income and earnings among people in the market; others narrow differences. Also, many government programs affect inequality more directly through, for example, tax deductions, food stamps, social security, Medicare, and corporate subsidies. Later in this book, we will look closely at the various initiatives Americans have taken, or chosen not to take, that shape inequality.

To see how policies also affect which particular individuals get to the top and which fall to the bottom of our ladder (i.e., the equality of opportunity), consider these examples: The amount of schooling young Americans receive heavily determines the jobs they get and the income they make. In turn, educational policies—what sorts of schools are provided, the way school resources are distributed (usually according to the community in which children live), teaching methods such as tracking, and so on—strongly affect how much schooling children receive. Similarly, local employment opportunities constrain how well people can do economically. Whether and where governments promote jobs or fail to do so will, in turn, influence who is poised for well-paid employment and who is not.

Claiming that intentional policies have significantly constructed the inequalities we have and that other policies could change those inequalities may seem a novel idea in the current ideological climate. So many voices tell us that inequality is the result of individuals' "natural" talents in a "natural" market. Nature defeats any sentimental efforts by society to reduce inequality, they say; such efforts should therefore be dropped as futile and wasteful. Appeals to nature are common and comforting. As Kenneth Bock wrote in his study of social philosophy, "We have been quick to seek explanations of our problems and failures in what we *are* instead of what we *do*. We seem wedded to the belief that our situation is a consequence of our nature rather than of our historical acts."[7] In this case, appeals to nature are shortsighted.

Arguments from nature are useless for answering the question of what determines the structure of rewards because that question concerns differences in equality *among societies*. Theories of natural inequality cannot tell us why countries with such similar genetic stocks (and economic markets) as the United States, Canada, England, and Sweden can vary so much in the degree of economic inequality their citizens experience. The answer lies in deliberate policies.

Appeals to nature also cannot satisfactorily answer even the first question: Why do some *individuals* get ahead and some fall behind? Certainly, genetic endowment helps. Being tall, slender, good-looking, healthy, male, and white helps in the race for success, and these traits are totally or partly determined genetically. But these traits matter to the degree that society makes them matter—determining how much, for example, good looks or white skin are rewarded. More important yet than these traits are the social milieux in which people grow up and live.

9

Realizing that intentional policies account for much of our expanding inequality is not only more accurate than theories of natural inequality; it is also more optimistic. We are today more unequal than we have been in seventy years. We are more unequal than any other affluent Western nation. Intentional policies could change those conditions, could reduce and reverse our rush to a polarized society, could bring us closer to the average inequality in the West, could expand both equality of opportunity and equality of result.

Still, the "natural inequality" viewpoint is a popular one. Unequal outcomes, the best-selling *Bell Curve* argues, are the returns from a fair process that sorts people out according to how intelligent they are. But *The Bell Curve*'s explanation of inequality is inadequate. The authors err in assuming that human talents can be reduced to a single, fixed, and essentially innate skill they label intelligence. They err in asserting that this trait largely determines how people end up in life. And they err in imagining that individual competition explains the structure of inequality in society. In this book, we use *The Bell Curve* as a starting point for really understanding inequality in America. By exploring that book's argument and its evidence, we can see what is wrong with the viewpoint that inequality is fated by nature and see instead how social milieux and social policy create inequality.

Generations of social scientists have studied inequality. Hundreds of books and articles have appeared in the last decade alone examining the many factors that affect who gets ahead and who falls behind in our society, including among those factors intelligence. We will draw on this treasury of research. We will also show, using the very same survey used in *The Bell Curve*, that social environment is more important in helping determine which American becomes poor than is "native intelligence" most generously estimated. Then, we will turn to the more profound question, the second question, of why the United States has the system of inequality it does. We will show that although some inequality results from market forces, much of it—and even many aspects of market inequality itself—results from purposeful, and alterable, policy.

THE BELL CURVE CONTROVERSY

In late 1994 a publishing sensation burst upon America. The covers of newsmagazines heralded a new study—perhaps the definitive study, the articles inside suggested—of the differences in intelligence between blacks and whites in America. *The New Republic* blared "Race & IQ" in enormous letters—and sold out all rack copies in Harvard Square. *Newsweek*'s cover featured facial profiles of a black man and a white man standing back-to-back with the superimposed words "IQ. Is It Destiny? A Hard Look at a Controversial New Book on Race, Class & Success."

Those who went beyond the front covers read of a book claiming that blacks are not as smart as whites, most likely because the two groups' genes differ. More broadly, they read that intelligence is a gift distributed by nature unequally at conception and that this distribution explains the inequalities among Americans. The political implications were clear: If inequality is natural, then governmental intervention to moderate it is at best wrongheaded and at worst destructive.

The book was attacked even as it was publicized. Both *The New Republic* and *Newsweek* bracketed their reports with critical sidebars, over a dozen in the first case; the *New York Times Magazine* published a cover-story profile of one of the authors implying that he is a boor; an interviewer for National Public Radio delivered almost every question to that author with a clear note of skepticism; the *New York Times* published at least two editorials against the book; and so on. And yet the book withstood the attacks and sold hundreds of thousands of hardcover copies (perhaps a sales record for a book with dozens of pages of statistical tables).

The Bell Curve, by Richard J. Herrnstein (who had long been a psychology professor at Harvard University at his untimely death shortly before the book's publication) and Charles Murray (a Ph.D. in political science, well-known conservative essayist, and resident at conservative think tanks), is more substantial than its media representations suggest. Its substance is due not merely to its mass, about 850 pages cover-to-cover, nor to its imposing array of graphs, tables, footnotes, and references. At its base is a philosophy ages old: *Human misery is natural and beyond human redemption; inequality is fated; and people deserve, by virtue of their native talents, the positions they have in society.* From that ideological base, Herrnstein and Murray build a case that critics cannot simply dismiss out of hand.

Herrnstein and Murray argue—relying on their own analysis of a large national survey, supplemented by an array of citations—that individuals' intelligence largely decides their life outcomes. Intelligence is distributed unequally among people, in a distribution shaped like a "bell curve" with a few people at the lower end, a few people at the upper end, and most people clumped in the middle. A person's position in that distribution heavily influences his or her position in the other distributions of life—the distributions of jobs, income, marriage, criminality, and the like.

The centerpiece of Herrnstein and Murray's evidence is the National

Longitudinal Survey of Youth (NLSY), a massive survey of over ten thousand young Americans involving repeated interviews over more than a decade. The NLSY administered the Armed Forces Qualifying Test (AFQT) to its subjects in 1980. Herrnstein and Murray show that NLSY subjects who scored high on that test, which the authors treat as an "IQ" test, were usually doing well ten years later, and those who had low scores ended up poorly. This is proof, they argue, that intelligence largely determines life outcomes. Herrnstein and Murray also contend that intelligence is essentially fixed, unchangeable in any significant fashion. People's fates are therefore also unchangeable. And so must be social inequality itself. Efforts to alter this naturally unequal order waste money and undermine its efficiency and justice. (Appendix 1 summarizes *The Bell Curve* in detail for those who have not read it.)

The Bell Curve is an inadequate explanation of where individual Americans end up in the system of inequality. Its answer to the question of why some people end up higher than others on the ladder of success vastly overestimates the relative importance of aptitude tests and underestimates the importance of the social environment. Despite Herrnstein and Murray's self-congratulations that, in examining intelligence, they have dared to go where no social scientist has gone before, scholars long ago established that scores on IQ and IQ-like tests were only of modest importance compared with social context in explaining individual attainment. We reinforce and expand that familiar conclusion by redoing Herrnstein and Murray's analysis of the NLSY survey. We show that they made major errors that exaggerated the role of the AFQT relative to social factors. For example, the AFQT is largely a measure of *instruction*, not native intelligence. (The *Newsweek* cover could just as well read "Grades. Are They Destiny?") Moreover, a correct analysis of the NLSY survey reveals that the AFQT score is only one factor among several that predict how well people do; of these factors, the social ones are more important than the test score.

More fundamentally, *The Bell Curve* also provides an inadequate understanding of systems of inequality. Its implied answer to the question of why the American ladder is so tall and narrow is that natural talent prevails in a natural market. This interpretation is wrong, in part because it is historically naive. For example, during most of this century Americans became substantially more equal economically, but since 1973 they have become substantially less equal. Understanding such fluctuations in inequality requires a broader historical and international perspective than *The Bell Curve* provides. We try to provide such a broader perspective.

Why do we pay so much attention to *The Bell Curve*? Some colleagues told us that *The Bell Curve* is so patently wrongheaded that it would be quickly dismissed; that genetic explanations of inequality are old news, having gained notoriety and disrepute thirty years ago, seventy years ago, and earlier; that we would only further publicize *The Bell Curve*; and so forth. But we felt that *The Bell Curve* is not easily ignored. It will not go away. Its ideas and data, at least as transmitted by the media and by politicians, will provide a touchstone in policy debates for many years. Our Berkeley colleague Troy Duster notes that within weeks of *The Bell Curve*'s publication, Charles Murray had been invited to address the newly elected Republicans in the House of Representatives and that an article in *The Chronicle of Philanthropy* had speculated that charity for "people of lesser ability" might be a waste.[8] Shortly afterward, the president of Rutgers University faced an uproar when he apparently alluded to *The Bell Curve* in explaining problems of black students.

Also, *The Bell Curve*'s perspective on society, which reduces a complex reality to little more than a footrace among unequally swift individuals, offends us as social scientists. *Social* reality—for example, how societies set up the "race" and how they reward the runners—cannot be understood through such reductionist thinking.

Nor were we satisfied with the critical appraisals that had appeared when we undertook this project in late 1994. Some reviewers, even as they castigated *The Bell Curve*, accepted, or were perhaps intimidated by, its scientific presentation. Some attacked the authors, the authors' funders, or the authors' intellectual friends. Deserved or not, such attacks do not invalidate Herrnstein and Murray's claims. Some commentators seemed to be grasping at straws, picking one or two contrary studies reported in the book without noting that the authors had piled on many others to support their arguments. And some just admonished *The Bell Curve* for its political implications. We believed that the book deserved neither the deference nor the unfair attacks. It could be challenged on scientific grounds. Also, in responding, critics generally accepted Herrnstein and Murray's framing of the question: why some people finish first and others last.[9] We do not.

As academics, we have the impulse to contest every claim and statistic in the 850 pages of *The Bell Curve*. There are certainly many errors and contradictions in the details.[10] However, there are more basic issues to address: What is intelligence? What role do individual talent and social environment play in shaping life outcomes? Why is the structure of outcomes set as it is? What difference does policy make? For resolving many of these issues, the particular statistics usually do not matter as much as logic and history. We will show that *The Bell Curve* is wrong statistically, that it is

even more profoundly wrong logically and historically, and that its implications are destructive.

One statistic is worth noting right away because it shows that there is
less to *The Bell Curve* than some intimidated reviewers have realized: "explained variance." Near the end of their text, Herrnstein and Murray capsu-
lize their argument by asserting that "intelligence has a powerful bearing
on how people do in life" (p. 527). However, 410 pages earlier they admit
that AFQT scores, their measure of IQ, explain "usually less than ten percent and often less than five percent" of the variance in how people do in
life (p. 117). What does "explained variance" mean? It refers to the amount
of the variation in some outcome, like income, from zero to 100 percent,
that can be explained by a particular cause or set of causes. To state that
intelligence explains 10 percent of the variance in, say, people's earnings
is to say that intelligence accounts for 10 percent of the differences among
people in earnings, leaving 90 percent of the differences among earners
unaccounted for. By Herrnstein and Murray's own statistical estimate, only
5 to 10 percent of the differences in life outcomes among respondents—the
odds that they became poor, criminal, unwed mothers, and so on—can be
accounted for by differences among them in AFQT scores. Put another
way, if we could magically give everyone identical IQs, we would still see
90 to 95 percent of the inequality we see today. What that means is shown
graphically in figure 1.2.

The figure displays the distribution of household income in the United
States in 1993. Across the bottom are the incomes from zero to $150,000.
The *Y*-axis represents the proportion of American households. The solid
line shows that virtually no households had zero income in 1993; about
.02—that is, 2 percent—had incomes of $25,000, about .01 (1 percent) had
incomes of $75,000; and so on. The solid line displays the actual distribu-
tion or the *shape of inequality* in household income. The dashed line displays what that distribution would have looked like if every adult in the
United States had had identical intelligence as measured by the AFQT:
hardly changed. Because AFQT score accounts for, at best, only 10 percent
of the variation in earnings, it leaves 90 percent of the variation unaccounted for.[11] In sum, intelligence, at least as measured by the AFQT, is of
such minor importance that American income inequality would hardly
change even if everyone had the same AFQT score. (In a response to similar criticisms, Murray backed away from explained variance as a criterion
for judging the importance of intelligence, but *The Bell Curve* argument
depends on that criterion.)[12]

As some economists have noted in reviewing *The Bell Curve*,[13] the issue
for policy is neither total explained variance nor even whether it is intelli-

If all adults had the same test scores (but different family
origins and environments), inequality of household
incomes would decrease by about 10 percent.

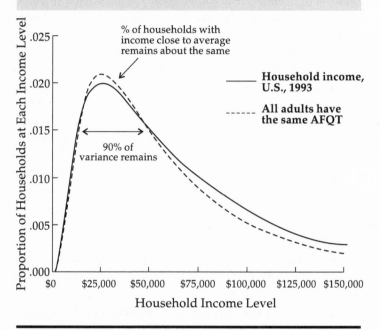

1.2. Explained Variance in Household Income Accounted for by Intelligence (*Note*: See text and notes for method of calculation)

gence or the social environment that explains more of the variation in individual outcomes. It is whether a given intervention can make a positive net
contribution to outcomes. In asserting that cognitive ability is critical to
determining individuals' fortunes but is unchangeable, Herrnstein and
Murray argue that no intervention can pay off. We will see, however, that
cognitive abilities are malleable (chapters 2 and 7). In asserting that socioeconomic background is of trivial importance in influencing individual outcomes, Herrnstein and Murray are claiming that working on social
conditions cannot be effective. We will see, however, that socioeconomic
conditions matter a great deal, so that policy there can be effective in increasing opportunity (see chapter 4). More important yet, Herrnstein and
Murray do not consider the deeper ways that social policy shapes both
individual competition and the structure of inequality (see chapters 5 and
6). There is great leverage for policy there, as well.

The claim that intelligence accounts for individuals' locations on the ladder of inequality is the central argument of *The Bell Curve*. But many discussions in *The Bell Curve* wander from that argument. The major such distraction is the discussion of ethnicity and IQ. It is a distraction because the argument over intelligence and inequality is unchanged whether or not there are inherent racial differences in intelligence. Charles Murray has admitted that, in the end, whether genes or environment explain racial differences in IQ scores *"doesn't much matter"* (italics in original).[14] We agree (although the genes versus environment debate matters a great deal if we want to explain racial differences in life circumstances). Because the media featured the topic of race and IQ so centrally, we must address the issue (see chapter 8). But otherwise we intend to stay on the main line of discussion. Finally, we agree with Herrnstein and Murray on some matters. (Secondhand readers of *The Bell Curve* may be surprised to learn that in some ways Herrnstein and Murray are not always conservative in their policy suggestions.) For example, we agree with them that Americans have since 1970 become increasingly polarized between rich and poor, and we agree with them that a guaranteed annual income ought to be considered as a possible national policy.[15]

We raise several arguments against *The Bell Curve*, any one of which is sufficient to dismiss it. If intelligence is not single, unitary, and fixed; if intelligence can be altered; if test scores mismeasure intelligence; if intelligence is not the major cause of people's fortunes; if markets do not fairly reward intelligence; if patterns of inequality are socially constructed—if any of these arguments holds, *The Bell Curve* case fails.

In the end, we respond in detail to *The Bell Curve* because it affords us an opportunity to explain what *does* account for the inequality we see in America today. That explanation stresses the importance of social environment and of policies that construct the social environment. That understanding, in turn, begins a realistic discussion of how to reduce inequality and its harmful effects.

OVERVIEW OF THE ARGUMENT

If one asks why some people get ahead and some people fall behind, answers concerning natural differences in ability are woefully inadequate. We can see that by looking closely at "intelligence." One reason inequality in intelligence is a poor explanation of class inequality is that individuals' abilities are much more complex, variable, and changeable than is sug-

16

gested by the old-fashioned notions of intelligence upon which *The Bell Curve* rests. Concretely, the basic measure of intelligence that Herrnstein and Murray use, the AFQT, is actually not a test of genetic capacity or of quick-wittedness. It is instead a test of what people have been taught, especially in high school, of how much they recall, and of how much effort they make in the test. Another reason that intelligence is not an adequate explanation of individual success or failure is that, as social scientists have known for decades, intelligence as measured by such tests is only one among many factors that affect individuals' success or failure. In the NLSY, respondents' AFQT scores in 1980 do not explain well how they ended up at the end of the 1980s. We show that, instead, aspects of respondents' social environments explain the outcomes more fully.

If one asks the more basic question of what determines the pattern of inequality, answers concerning individual intelligence are largely irrelevant. Societies and historical epochs vary greatly in the nature and degree of their inequality; they differ much more than any variations in intelligence, or the market, can account for. Some of that variability lies in technological, economic, and cultural changes. But much of it lies in specific policies concerning matters such as schooling, jobs, and taxes.

In the end, we *can* change inequality. We *have* changed inequality. American policies have reduced inequality in many spheres—for example, improving the economic fortunes of the elderly—and have expanded inequality in others—for example, with tax expenditures that advantage many of the already advantaged. And the experience of other nations shows that there is much more that can be done to reduce inequality if we choose to do so.

Policies also affect where individuals end up on the ladder of inequality. Policies help construct social environments. Policies even alter cognitive skills, particularly in the ways we structure schooling. The leverage here lies not with the episodic compensatory programs over which there has been much debate, but with the everyday structure of schools in America.

Finally, what about race? Arguments that African Americans and Latino Americans have done poorly in the United States because they are less intelligent than whites are completely backward. The experiences of low-caste groups around the world show that subordinate ethnic minorities do worse in schools and on school tests than do dominant groups, whatever the genetic differences or similarities between them. Whether it is Eastern European Jews in 1910 New York, the Irish in England, Koreans in Japan, or Afrikaaners in South Africa, being of lower caste or status makes people seem "dumb." The particular history of blacks and Mexicans in the United States fits the general pattern. *It is not that low intelligence leads to inferior status; it is that inferior status leads to low intelligence test scores.*

THE PLAN OF THIS BOOK

Chapter 2 examines how the psychometricians upon whom Herrnstein and Murray draw for their psychology have sought to study "intelligence." The psychometric concept arises largely from the IQ tests themselves: Intelligence is the statistical core (labeled "*g*") of those tests. In other words, intelligence is what IQ tests measure. We show how problematic that definition is by showing that the AFQT largely measures how much math and English curricula teenagers have learned and display. But there are other, better ways to think about intelligence. We discuss as an example the "information-processing" perspective, one which is more realistic.

Chapter 3 examines *The Bell Curve*'s specific evidence about intelligence: scores on the AFQT. Scores on school achievement tests are, of course, important in a society that rewards people according to how well they do in school, but they are not what most people would consider as "intelligence" per se. We also explore the ways Herrnstein and Murray "massage" the AFQT data to fit their arguments. They overstate the validity and utility of the AFQT scores. Yet to the degree that such test scores measure how well we educate our children, their ability to predict life outcomes testifies to how critical educational policy is for American inequality.

Chapter 4 addresses the fattest section of *The Bell Curve*, its statistical analyses purporting to show that NLSY respondents' AFQT scores best predict—and so, presumably, the respondents' intelligence most determines—what becomes of them. We review critical errors Herrnstein and Murray made in their analysis; we reanalyze the identical data; and we come—as other scholars have, also—to opposite conclusions: Social environment is more, not less, important than test scores in explaining poverty, likelihood of incarceration, and likelihood of having a child out of wedlock. For economic outcomes, gender, a trait Herrnstein and Murray ignored, matters most of all. Other social factors—education and community conditions—are at least as important as test scores. Stepping back from the specific data, we point out that these findings are not news to social scientists. We have long understood that a person's economic fortunes are hostage to his or her gender, parents' assets, schooling, marital status, community's economy, stage in the business cycle, and so on; intelligence is just one item on such a list. This chapter settles the issue of why some people get ahead of others in the race for success; the next chapter looks at what the racers win or lose.

Chapter 5 turns attention to *systems* of inequality, showing how greatly they vary across history and among nations. The question of chapter 5 is not whether individuals are more or less equal, but whether *societies* are. We will see how the degree of inequality fluctuated in American history, particularly how inequalities widened since the 1970s. And we will see how extreme the United States is compared with other advanced industrial nations. The inequality in America today is not "natural" but in great measure the result of policies that tolerate wide inequalities. Ironically, those policies are, despite assertions by interested parties, *not* necessary for economic growth; indeed, inequality may well retard economic growth.

Chapter 6 turns to several explicit national policies that structure inequality in America. Some policies and programs narrow inequality—social security, Medicare, food stamps, etc.—while some widen it—corporate subsidies, the mortgage deduction, laws concerning unionization, and so on. Compared with America's economic competitors, we do relatively little to equalize people's economic fortunes—or even their economic opportunities. This explains our charge that the inequality Americans have is a result of the policies Americans have at least tacitly chosen.

Chapter 7 turns to policies that shape individual abilities, specifically, intelligence. Individuals' cognitive skills—those supposedly fixed talents that determine economic inequality—are indeed changeable. We show, using the examples of the school year, tracking in schools, and the structure of jobs, that learning environments alter how and how well people think. Policies help construct those learning environments. Even the inequality of ability is subject to social shaping.

Chapter 8 turns to race and ethnicity—a topic we believe was a distraction, albeit an incendiary one, in *The Bell Curve*. Why do blacks and Latinos score lower on standardized tests? This turns out to be not a biological question but a social one. Around the world, members of disadvantaged groups usually score lower than members of advantaged groups, whatever their racial identities. In many cases, both the higher- and lower-status groups are of the same race. Also hard to reconcile with the racialist viewpoint is the way ethnic groups seemingly become smarter *after* they have succeeded. For example, in Japan Koreans are "dull," while in the United States Koreans are "bright"; Jews in America were "dull" seventy-five years ago but are among the "cognitive elite" today. We describe three

19

ways that ethnic subordination in a caste or castelike system leads to poor school and test performance: One, subordination means material deprivation for students, which in turn impairs their achievement; two, subordination usually involves group segregation and concentration, which, by multiplying disadvantage and drawing all group members into difficult learning situations, undercuts academic achievement; and three, subordination produces a stigmatized identity of inferiority, which in turn breeds resignation or rebellion, both of which limit academic achievement. The histories of African Americans and Latino Americans, as well as their current conditions, more than suffice to explain why their members tend to score lower than whites on tests and also why they do less well in the race for success. The American case fits the global pattern; it is not genes but caste positions that explain the apparent differences in cognitive performance.

Chapter 9 concludes with a consideration of what the intellectual and the practical implications are of understanding inequality in these historical and sociological ways.

CONCLUDING COMMENTS BY WAY OF INTRODUCTION

A comment on the "burden of proof": Many readers, by now accustomed to contradictory studies about how certain foods do or do not cause heart disease or cancer, may feel unable to decide among dueling Ph.D.s' claims about inequality. In this book, we contest many specific issues of evidence in *The Bell Curve*. But more important is how the basic questions are framed and the historical breadth of evidence examined. From such a fundamental perspective, we find that intelligence, broadly understood, does affect Americans' fates but is just one factor among many. It is not the key to American inequality nor to American social problems; indeed, differences among individuals altogether are not the key. The key is how we, together as citizens, choose to structure our society. We do not, of course, have unlimited freedom of action; we are constrained by material circumstances, social traditions, and political institutions. But we have a lot more freedom to act, this will book will show, than admitted by those who counsel acceptance of the growing inequalities in our society. The challenge is to make those choices.

In thinking about those choices, it may help to go back to first principles. This nation draws its moral precepts from its biblical and republican traditions. The Bible repeatedly enjoins us to help the needy; the Declaration of

Independence announces that "all men are created equal, that they are endowed by their Creator with certain inalienable Rights."[16] Such a nation should presume that its people come equally equipped to fulfill those promises. The burden is on those who would contend otherwise, who would have us sorted out at birth into the worthy and the unworthy. The burden of proof is on those who would contend that some of us are hopeless and fated only for piteous charity. Absent conclusive proof of that claim, Americans should assume an equality of worth and move to expanding every American's horizon.

Understanding "Intelligence"

> "When *I* use a word," Humpty Dumpty said, in a rather scornful
> tone, "it means just what I choose it to mean—neither more nor
> less."
> "The question is," said Alice, "whether you *can* make words
> mean so many different things."
> "The question is," said Humpty Dumpty, "which is to be mas-
> ter—that's all."
> —Lewis Carroll, *Through the Looking Glass*

THE WORD here is "intelligence." For those who believe that inequality
of talent explains inequality of fortunes, intelligence is the most important
talent of them all. For those who believe that talent is largely fixed at birth,
the *immutability* of intelligence dooms any effort to alter inequality of outcomes. But arguments that intelligence is immutable rest on a particular
way of understanding intelligence, an approach called psychometrics (the
measurement of mental traits). In this chapter, we will show that this outdated approach underestimates how complex and flexible people's cognitive skills are and thereby underestimates how much such skills can be
improved.

Other, newer schools of psychology offer better approaches to understanding intelligence and also offer sensible hope for improving Americans' abilities. They show us that we *can* beneficially invest more in educating people, young and old; that we can train people to use their minds
more acutely, to make more of their abilities; and that we can produce a
more intelligent population—just as the establishment of mass education
did generations ago. We will take, as one example of such alternatives to
psychometrics, the *information-processing* perspective. It encompasses
psychometrics but extends our knowledge of intelligence much further.
Contrasting psychological schools is not a mere academic exercise. The
differences are akin to the difference between seeing the moon as a fixed
orb in the sky and seeing it as an orbiting companion of Earth. If you believe that the moon is fixed, you could never land on it because you would
always aim at the wrong place. If you realize that it is in motion, the moon
can become the point of departure for exploration of the cosmos. So with

understandings of intelligence: When we understand how malleable it is, we can target our efforts on expanding and enriching it. To the degree that people's cognitive skills affect their chances of success, then understanding how malleable those skills are underlines how much equality of opportunity can be changed.

In *The Bell Curve*, Herrnstein and Murray imply that no self-respecting psychologist questions the psychometric perspective. But not only do many established psychologists question it, most of the important insights into intelligence come from perspectives that directly contradict fundamental claims of psychometricians.[1] In this chapter, we describe the differences between the schools, showing how limited psychometrics is for understanding the role of intelligence in inequality.

Psychometrics, we will show, has not been centrally concerned with how people think or solve problems. It has been concerned instead with developing tools to rank, to *differentiate*, people by how successfully they solve academic problems. Psychometricians have spent decades refining minutely graded tests so that they yield scores for individuals that are reliable and that correlate with success on other tasks, such as progress in school and in the job market. The specific problems that comprise a test are less important to psychometricians than that the test "works" to rank people. Among the difficulties with this approach is that the test items psychometricians typically use deal with school subjects or are school-like, such as math questions. They thereby confound intelligence with schooling. (We will see this in detail below.) Another profound difficulty is that the psychometric effort to rank people requires that the tests produce fine distinctions, differences in scores that correspond little to differences among people in how they use intelligence in everyday life. Yet another problem is the psychometricians' insistence—driven by their techniques rather than their observations of people—that there is *a* singular and basic intelligence. Other approaches to intelligence avoid these and similar difficulties, and they do so by studying thinking itself, typically in ways too complex to be easily absorbed into psychometric tests.

Before fleshing out these points, we need to understand how progress is made in science. Otherwise, the psychometric school and its alternatives will simply appear to be equally valid points of view. Therefore, we first discuss how one research framework in science gives way to another and why. Second, we examine psychometrics and demonstrate its fundamental circularity: Psychometricians have discovered a single, fixed "intelligence" *because* they have developed a methodology that works, and that methodology works because it implicitly assumes that there is a single intelligence, essentially fixed at birth or early in life. Third, we point to alternative ways of understanding intelligence, looking in particular at the information-processing framework, which understands intelligence to be changeable.

THE IMPORTANCE OF "PARADIGMS"

The stereotype of the scientist as a lonely inquisitive soul is false.[2] Without communication among scholars, each one would be forced to start from scratch, and accumulating knowledge would be impossible. To communicate, researchers use a common "language" of concepts—a *paradigm*. Like any language, paradigms impose limitations. Crucially, paradigms limit the questions that researchers can ask. Some research questions are critical to a paradigm and others make no sense in that paradigm. For example, in the Flat-Earth paradigm, questions about what happens if you sail off the edge of the Earth make sense, while questions about the velocity needed to reach orbit are nonsense. In the Spherical-Earth paradigm, the reverse is true. Because the list of all possible questions is infinite, we need paradigms to delimit our inquiries. Paradigms also alert researchers to what the answers to those questions should look like.

One way paradigms limit questions and answers is by defining key concepts; another is by suggesting how to observe, or measure, those concepts. Like the questions, empirical observations typically make sense only *within* a paradigm. For example, precise measurements of bumps on people's heads made sense for the nineteenth-century "science" of phrenology but make no sense to twentieth-century psychologists. It is a paradigm that gives any particular observation its meaning. We will see that IQ test scores have scientific meaning only within the perspective, or paradigm, of psychometrics. (They have, of course, considerable practical meaning as well.)

New paradigms arise when old paradigms lose steam. One way a paradigm loses steam is by running out of stimulating questions; younger scholars find it boring and "tired." Another way is that scholars raise new questions that cannot be addressed in the old paradigm. A third route to paradigm change is that new research findings appear that cannot be assimilated by the old paradigm. (We will discuss an example in psychometrics—the phenomenal rise in IQ scores over the twentieth century—later.) When apparent anomalies arise, scholars first try to patch up the old paradigm in an ad hoc way. Eventually, so many ad hoc patches are applied that it no longer seems to be an elegant blueprint for research but instead seems to be a hodgepodge of poorly understood findings. This is roughly what happened to the astronomical paradigm that assumed that the Sun and planets revolved around Earth; it became so complicated trying to account for planetary motion that the paradigm fell apart. The version of psychometrics upon which *The Bell Curve* is based is, we suggest, in such a state.

A perhaps fanciful story may illustrate the power paradigms have to shape questions and observations. It deals with the concept of human height. Suppose that there exists an isolated community of people who commonly attend public events such as concerts, plays, and speeches. These performances are given at flat outdoor plazas, the performers are at ground level, and the members of the audience stand throughout each performance. Among these people, "height" is a critical concept. They come to define height as the vertical length of a standing human being, plus whatever hair, headgear, high heels, or other footwear the person is wearing. A critical question people ask when planning to attend a concert, speech, or play is "How tall will you be?" It is an important question in this society because answers to it tell people where they should stand.

One day, an enterprising (and tired) young woman fashions a small bench to bring to the gathering, and, sitting in the front row of the plaza, she amazes many at her reduced "height." Others soon see the gain in sitting, and eventually chairs abound. Upon making their way to the plaza, many persist in asking the old question, "How tall will you be?" The answers, however, are now very confusing. Some people cling to the old concept and report the total height of their body and accoutrements. Others stare blankly at the question, as they know not what to report given the chair in their hands. The old concept no longer fits the new reality.

Eventually, arguments ensue. Some alter the concept of human height to mean the vertical distance that a person takes up whether sitting or standing. Others argue that the concept is useless and call for society to dispense with it. Still others try to resurrect the concept by correlating the old measures of height to the vertical distance of people sitting at the plaza. Finding a high correlation between these two measures, the call goes forth to hold on to the old ways of thinking.

As we leave this community to its problem, we may draw a few lessons: (1) The concept (height) arose from important problems in that society; (2) reality (i.e., the dimensions of space) limited the definitions of the concept that could be devised, but it did not determine a unique definition; (3) the old concept became problematic because it no longer answered the questions that society needed answered; and (4) the old concept could be fit under a new framework but would likely not have been central to that new framework. The basic point is that understandings of "height" rested on the prior assumptions; it could have different meanings and different measurements. So, too, with intelligence.

We will focus on two paradigms for thinking about intelligence: psychometrics and information processing. We will see how each defines "intelligence," how practitioners try to measure intelligence, and what their definition implies about policy toward intelligence. Because it is trapped in a narrowly framed perspective on intelligence, psychometrics can provide only a fatalistic view of the opportunities for change. It should become apparent that newer approaches, such as information processing, allow us to break out of the dated confines of psychometrics, allowing us to see how we can raise intelligence and modify inequality.

INTELLIGENCE IN THE PSYCHOMETRIC SCHOOL

The psychometric paradigm assumes that the fundamental skill or talent critical to human functioning is "intelligence." It gives no or little attention to determination, self-discipline, empathy, creativity, charm, energy, or a myriad of human abilities that we recognize in people during our daily lives. Occasionally, a psychometrician suggests (as Herrnstein and Murray do) that other admirable traits result *from* intelligence. Famed psychometrician Edward Thorndike once wrote that "the abler persons in the world . . . are the more clean, decent, just, and kind."[3] But for the most part *the* talent of interest is intelligence.

Psychometricians' descriptions of intelligence arose from their efforts to sort and rank people. Among the foundational, paradigmatic assumptions of psychometrics is the premise that people *must* rank in a "bell curve." Assumptions such as this one severely limit the approach, rendering psychometrics at best irrelevant and at worst a hindrance to understanding inequality. Let us see why this is so.

In *The Bell Curve*, Herrnstein and Murray assume that there is "a *general capacity* for inferring and applying relationships drawn from experience," which is synonymous with a "person's capacity for complex mental work" (p. 4; emphasis added). This basic, singular capacity is what psychometricians label g, or general intelligence. Herrnstein and Murray contend that psychometricians have accepted the reality of g, but even a cursory scan of the psychometric literature reveals contentious disagreement among them about the existence or importance of g. Psychometric positions range from the idea of a unitary g, through Raymond Cattell's distinction between

---

THE HISTORY OF TESTING AND THE LIMITS OF THAT HISTORY

Critics of psychometrics often stress the political history of IQ testing in the United States and the ideological commitments of many psychometricians.* They highlight the use of IQ tests in the early to mid-twentieth century to promote eugenics, restrict immigration, and defend segregation. (Several prominent psychometricians explicitly allied themselves with coarse racists and Nazi sympathizers.) Based on this history, many dismiss testing.

We do not take this approach, for two reasons. First, the history of psychometric testing in other countries shows that it is really a history of the *decisions* people made about how to use the tests. In the United States, IQ tests were used for discriminatory purposes, but elsewhere, like Great Britain, liberal reformers used similar tests to find promising lower-class children and provide them with opportunities.** Second, IQ and similar tests are widely used by educators and employers as gatekeeping mechanisms to determine who will or will not obtain scarce positions. It is highly unlikely that such tests will be discarded simply because people used them badly in the past. Therefore, we need instead to examine the logic of intelligence testing and its limitations.

* See, for example, Huston, *Testing Testing*; Kamin, "The Pioneers of IQ Testing"; Tucker, *The Science and Politics of Racial Research*.

** See Wooldridge, "Bell Curve Liberals."

---

crystallized and fluid intelligence, all the way to Joy Guilford's notion of 120 components of intelligence. No less prominent a psychometrician than Arthur Jensen, educational psychologist at Berkeley, disagrees that *g* can be boiled down to one entity; he suggests that psychometric *g* may reflect as many as four independent components.[4]

We review the chain of reasoning that leads to the psychometric notion of intelligence. We will see that, in practice, the definition of intelligence as "a general capacity for inferring and applying relationships" does not match the procedures psychometricians actually use. The tests, therefore, are not good measures of what most people would consider to be "intelligence."

We begin with Arthur Jensen's definition: *"intelligence is what intelligence tests measure"* (emphasis added).[5] This is an honest and telling statement. In practice, psychometricians have defined intelligence *after the fact*: after constructing intelligence tests, obtaining the results, and interpreting what those results mean.[6] So, what then is an intelligence test?

Measuring Intelligence in the Psychometric Tradition

For psychometricians, virtually any task can be an intelligence test because intelligence presumably determines success on virtually all tasks. But some tests are better than others, and the best intelligence tests have certain properties. These properties are: (1) specificity; (2) stability; and (3) differentiability. The first two are properties that any good measure of anything should have. It is the third one, differentiability, that is problematic.

Specificity means that a good test will test only one kind of skill. If, for example, we want to test people's strength, we would not want to use a test that also reflected their agility.[7] To be well constructed, therefore, an intelligence test must reflect intelligence only, and it must not capture other potentially important determinants of task success, such as motivation, creativity, and anxiety. (This will turn out to be a problem for the test used in *The Bell Curve*.) *Stability* means that if we administer the same test twice, the measure should not change. We would be puzzled if a football player lifted 500 pounds one day and was unable to do so five days later. We know that the measuring instrument, the 500-pound barbell, has not changed, which is another way of saying that lifting the barbell is a stable test. Similarly, repeated applications of an intelligence test should yield a consistent score for each test taker unless that test taker's intelligence changes in the interim. (In reality, scores on many common tests of cognitive skills, such as the college Scholastic Assessment Test [SAT], reflect a notable contribution of randomness. That is why modest differences in such test scores—for example, sixty points on the verbal SAT—should not be considered definitive.)[8] Specificity and stability are good properties for all measuring instruments, including yardsticks, thermometers, altimeters, and more.

The third property of a well-constructed psychometric test, *differentiability*, is not necessarily a good property of a measuring instrument. For example, we would not throw away a tape measure because it cannot help us discern tiny differences in height among prospective football players. Each player has a specific and discernible height, but quarter-inch differences are not relevant. Yet, if we *require* that a test differentiate to the finest degree, then we must throw away measures that do not do so, like our tape measure, and search for more precise ones. Psychometric tests drafted to

---

DRIVING TESTS AND PSYCHOMETRIC TESTS

Many tests are designed to certify sufficient competence rather than to differentiate the population. Driving tests are an important example. We certify sufficient competence, which has allowed the vast majority of drivers to use the highways. Alternatively, we could require the driver's license test to differentiate the population of would-be drivers in the same way that psychometric tests differentiate the population by intelligence. If we did so, such important skills as making a left turn in heavy traffic, correctly interpreting street signs, and parallel parking—skills that compose most driving tests—would be deemphasized or eventually removed from the test because they fail to make sharp distinctions among the population. Instead, ability to maintain control of the car at Indianapolis Speedway speeds, to make Grand Prix hairpin turns effectively, and to navigate a slalom course within a given time limit without damaging the pylons would form the driving test.

---

differentiate the population finely may take a difference that is comparable to that quarter-inch and magnify it—even if such fine distinctions have no practical implications (see box above).

Differentiating to a fine degree is not a need of testing. It is a *paradigmatic commitment* of the psychometric test designers. That is, psychometricians need to make fine distinctions in their tests because they presume that fine distinctions in intelligence are real and that they matter.

To understand that paradigmatic commitment, let us start with the very name "psychometrics." In scientific disciplines, such as biology and sociology, the suffix "-logy" means "the science of." The "-metrics" suffix refers to measurement. Since the 1930s most psychometricians have devoted their efforts to the measurement of intelligence. Their approach to this task may seem quite alien to the outsider. They say, in effect, let us *assume* the fundamentals—how much intelligence is out there in the population and how it is distributed among people—and set about finding each person's place on that assumed distribution. Why do they approach the task from this direction? Why do they not measure intelligence directly as researchers measure, say, educational attainment or income? The answer is in the units of measurement. Educational attainment and income have clear and measurable units—years and dollars. But nobody knows what *units of intelligence* are or how to count them. We do not have "ounces" or "watts" of intelligence.

Psychometricians have "solved" the problems caused by the lack of a unit of intelligence by assuming those problems away. They note (correctly) that *if* we assume (1) that intelligence in the population is distributed normally, like a "bell curve," and (2) that people can be ranked from the lowest to the highest on a test, then we can give each person an intelligence score by simply converting his or her rank into a score. *If* the previous two assumptions are correct, then the units of measurement turn out to be irrelevant; we do not actually need to see or weigh an amp or decibel of intelligence. Those are some big ifs. There is a basic circularity here. The scores people receive on intelligence tests derive from the measuring tools rather than from observations of intelligence at work.

What is the distribution of intelligence in the population? Psychometricians do not know. They *assume* that it is a bell curve. The bell curve is characterized by the large bulge in the middle representing the preponderance of scores near the average and by long, sloping edges indicating that large departures from average are infrequent. And the bell curve is perfectly symmetrical. But this distribution is not a discovery based on evidence about human intelligence. The first psychometricians *chose* the bell curve to represent the distribution of intelligence as a matter of faith that just about everything is distributed this way, and as a statistical convenience.[9] For technical reasons, it is easier to convert test scores into IQ scores using a "normal," bell-curve distribution than using any other statistical distribution.[10]

If a psychometrician assumes a bell curve and ranks people from first to last on a test, assigning scores is routine—as long as not too many people are tied at the top or at the bottom. And here is where differentiation becomes crucial. A bell curve, by definition, has very few people at the top and the bottom. Therefore, a "good" test must also have this property. The test must produce scores in a bell curve distribution, with very few people getting top scores and very few people getting bottom scores. Psychometricians routinely include questions in their tests to produce this very result. If they are building a vocabulary test (a typical part of intelligence tests), they make sure to include some words that only one person out of 100 will not know and some words that only one person in 100 will know. Whether knowing these rare words—say, "snood" or "entremets"—has anything to do with intelligence as "a capacity for inferring or applying relationships drawn from experience" is, in effect, irrelevant. Those very common and those very rare words define the ends of the scale.

Psychometricians then end up with scores for each test taker. But because the units of measurement are arbitrary, the numbers are not fixed like inches on a carpenter's steel rule. They are more like pencil marks on a rubber band that can be stretched to fit between any two points. Give psychometricians a word to define (or another problem) known to 99 percent of the population to nail one end of the rubber band to and a word known to only 1 percent of the population to nail the other end of the band to, and they will be able to fill in the rest of the distribution, the bell curve.

In sum, psychometricians *assume* a single intelligence and that it is distributed among people like a bell curve, and so they build their method to yield a bell curve. There is no substantive reason to start from these assumptions.

The Bell Curve in The Bell Curve

By looking at *The Bell Curve* itself, we can see how distortions arise from the commitment to differentiate and the commitment to the bell curve. Recall that the basic evidence Herrnstein and Murray draw upon is a test, the Armed Forces Qualifying Test, given in 1980 to over 12,000 young people in the National Longitudinal Survey of Youth. The AFQT is composed of four sets of questions for a total of 105 test items. (These sets are interspersed among six others, totaling another 228 questions, to form the Armed Services Vocational Aptitude Battery.) For the moment, we will accept Herrnstein and Murray's claim that the 105 questions measure intelligence. Later in the chapter, we will look at what those questions actually measure. If one graphs the number of questions answered correctly by how many people answered that number of questions correctly, one obtains the shaded shape labeled "Original" in figure 2.1. That shape, leaning heavily to the right, shows how many of the white test takers answered that many questions correctly. (To be consistent with Herrnstein and Murray's analysis of intelligence and class inequality, we largely analyze only the whites in the NLSY sample.) For example, 45 young adults answered 104 or 105 correctly and 126 answered fewer than 30 right. (Both lines in the figure represent "smoothed out" data; that is, slight and erratic variations in numbers are averaged out for legibility.)

Oops! Despite the best psychometric assumptions, this original distribution is not a bell curve! Most white respondents scored near the top end of

Originally, the distribution of test takers by the number of correct answers was bunched upward; it took considerable effort to transform it into a bell curve.

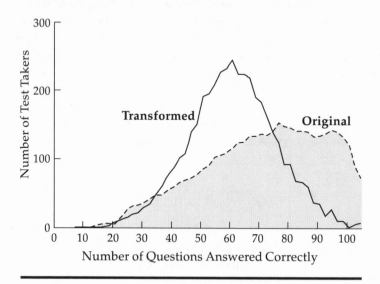

2.1. Distribution of Original Scores on the AFQT and Distribution of Scores as Transformed by Herrnstein and Murray (*Source*: Authors' analysis of the NLSY data)

the test. About 20 percent of the test takers answered more than 90 percent of the questions correctly. This is the real distribution of scores from the key measure in *The Bell Curve*. There are simple reasons why the AFQT did not yield a bell curve.[11] Our concern here, however, is with the psychometric insistence that there *must* be a bell curve. The other line in figure 2.1, labeled "Transformed," shows the distribution of test takers after Herrnstein and Murray recalculated the scores. It is roughly bell-shaped; it is also the source of the title and the jacket design for *The Bell Curve*. How did Herrnstein and Murray get a bell curve from the lopsided distribution of original scores? By a good deal of statistical mashing and stretching. Because they presumed, as psychometricians do, that intelligence must be distributed in a bell curve, they justified transforming the number of questions each test taker correctly answered until they produced the bell curve in the figure.[12]

(Practically, what Herrnstein and Murray did was give "extra credit" for being at the higher or lower ends. For example, a difference of three correct answers for those at the upper end, 105 versus 102, yielded differences in what Herrnstein and Murray labeled "zAFQT" scores of one full step, or "standard deviation," from 2 to 3. But for respondents near the middle, it took far more correct answers to move one step—28 more correct to move from −1 to 0. Near the bottom, it took 16 more correct answers to go from −3 to −2. The effect of this is to exaggerate, to stretch out, slight differences among test takers at the high and low ends in order to form a bell curve.)

If one builds a science around the axiom that an unseen entity, intelligence, must be distributed among people like height is, in a bell curve, one is forced into these kinds of contortions. Why does this matter? One reason it matters is that this transformation creates a bell curve where there was none. It shows how researchers can be trapped in a paradigm. Another is that this rescoring scheme exaggerates the importance of being at the highest and lowest ends. And that is what Herrnstein and Murray *wanted* to do. They "knew from collateral data that much of the important role of IQ occurs at the tails" (p. 573). So they did not use simple alternatives to that transformation, like centile score (99th percentile, 98th percentile, etc.) or raw score (number of questions correctly answered), but instead constructed the bell curve. By stretching out the tails, they helped affirm their expectations that the tails are critical. And the tails are critical to Herrnstein and Murray because their entire focus is on demonstrating that the top 5 percent (the "cognitive elite") do so much better and the bottom 5 percent (the "very dull") do so much worse than everyone else. For the most part, they ignore the middle 90 percent of Americans. This is not good science but self-fulfilling prophecy.[13]

In this way, the psychometric testing paradigm prefigures the meaning of intelligence. If "intelligence is what intelligence tests measure," then the problem is how to measure intelligence. The answer is to use a well-constructed intelligence test. How do we know a test is well-constructed? By whether it sharply differentiates the population. Hence, intelligence must be finely distributed (preferably in a bell curve). Herrnstein and Murray's analyses of the "cognitive elite" and the "very dull" would all be impossible without the presumption that intelligence is so finely graded that there are distinctive top and bottom 5 percents. And it makes sense only within the psychometric paradigm.

Psychometricians' methodological need to differentiate dovetails with other reasons to differentiate. Administrators, employers, and other consumers of psychometric tests use them to allocate, stratify, and label students and employees. Sorting people out into precise ranks may not be necessary for education or for work, but it is useful for deciding who to admit, hire, promote, and so forth. (It is as if we allowed only the top 5 percent of drivers to own sports cars. Then we would design driving tests to differentiate so finely. But that seems silly. When more precise distinctions are necessary, say for licensing heavy equipment drivers, then more precise tests are used.)

Tests constructed with differentiating as a goal also legitimate the stratification that is imposed, make it seem just. Tests are good devices of legitimation because they appear to be objective. The record on the use of intelligence tests to justify ethnic differences is a long one.[14] We will explore that role of tests in chapter 8.

Sharp differentiations make psychometric tests fundamentally unlike most other life tasks, like driving and doing one's job. A good test should be like a good task, and neither needs to differentiate in order to tell us about the talent of the person being tested. However, ability tests have not been used primarily to tell us about the talent of the person being tested; instead, they have been used to tell us about the talent of the person being tested in relation to the talent of other persons. The desire to compare performance, needed for statistical analysis and organizational work, requires that tests discern difference, no matter how trivial that difference may be in the larger scheme of things. Because we argue against a position that has become ingrained in the very idea of testing, we cannot state it strongly enough: It is possible to construct evaluations that point to the talents and weaknesses of people but that do *not* discover, magnify, and therefore solidify originally trivial differences. Such tests might incorporate evaluations of diverse skills in general ways.

Most skills are useful even if virtually everyone succeeds at them. Walking, for example, is no less useful because everyone learns to walk. An ability test that most people can pass equally well is, however, of little use to psychometricians or to many users of tests. Therefore, test makers behave as if every good test must differentiate. We argue, however, that a good test will behave like a good task, revealing whatever differences there are but not magnifying differences for the sake of differentiating people.

The Psychometric Circle

In these ways, psychometrics rests on statistics as much as on psychology. Its very definition of intelligence, g, derives from statistical finding that test takers' scores on various kinds of ability tests correlate with one another;

people who do well on one tend to do well on another. The explanation is that the tests all measure a common, underlying property: general intelligence, or g.[15] There is nothing wrong with beginning a research program trying to explain a statistical regularity such as the intercorrelation of tests, nor with making statistical regularity a goal, nor with setting aside non-statistical concerns while one is in the middle of this process. However, what is wrong is limiting the work to only these concerns. By closing off all other issues, the reasoning becomes irremediably circular.

To demonstrate this circularity we need to introduce one more concept used to evaluate tests: *validity*. A test has validity when it measures exactly what it is supposed to measure. The psychometric circularity is clear in how validity has been evaluated for intelligence tests. There are three main ways to establish validity, or three types of validity. Sometimes a measure appears valid in one way and invalid in others.

One type is *predictive* validity. A measure has predictive validity if it can predict an outcome that we assume ought be predicted by the underlying trait we are trying to measure. For example, if we assume that intelligent people get better college grades and we show that people who do well on the SAT also get better college grades, we can then say that the SAT appears to have predictive validity as a measure of intelligence, given that the outcome of interest is college grades. If we were to say that the SAT has predictive validity without stipulating what is being predicted, then our statement would be incomplete. To evaluate a claim of predictive validity, we need to know what is being predicted (grades) as well as what is the underlying concept being measured (intelligence).

Figure 2.2(a) illustrates these assumptions graphically. The double-headed arrow between grades and SAT score shows what the research uncovers, that the two are correlated. (A double-headed arrow indicates a correlation between two measures; a single-headed arrow indicates a correlation *and* an assumption about the direction of causality.) This exhausts all the empirical information that is available. Figures 2.2(b) and 2.2(c) show how analysts invoke additional assumptions to draw conclusions about what the SAT really measures.

Figure 2.2(b) illustrates how one may infer that the SAT has predictive validity as a test of intelligence when used to predict college grades. It assumes that intelligence determines performance on the SAT and that intelligence causes better college grades, so that the observed correlation between the SAT and college grades in figure 2.2(a) is the result of the unobserved entity, intelligence.

Predictive validity can imply different causes.

(a) Correlation

College Grades

SAT Scores

(b) Intelligence Interpretation

College Grades

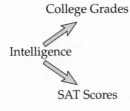

Intelligence

SAT Scores

(c) Social Class Interpretation

College Grades

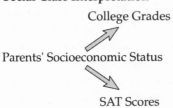

Parents' Socioeconomic Status

SAT Scores

2.2. Different Interpretations of Predictive Validity

The inference depends crucially on the two *assumed* causal relationships. They may be accurate, but we have no way of knowing for sure. To know for sure, we would have to have another, better measure of intelligence. Yet, if we had such a better measure, we would not need to investigate the predictive validity of the SAT as a measure of intelligence! In sum, the intelligence interpretation is just that, an interpretation of a finding, not a finding itself. Figure 2.2(c) makes that point more clearly. Here, we interpret the SAT as a measure of parents' socioeconomic status, believing that higher parental socioeconomic status determines both higher grades and higher scores on the SAT. Juxtapose figures 2.2(b) and 2.2(c) and you realize why predictive validity is a very weak form of validity. Taken together, they show that *any* concept that can be assumed to be related to both of the observed factors can be said to be captured by the test one is seeking to validate. It could have been height, body mass, hair color, interest in classical music, what have you.

Thus, predictive validity is a logically weak form of validity. Leaning on predictive validity is often a substitute for more research that would identify what a test really measures. But research psychometricians, after generations of work, continue to rely heavily on predictive validity.[16] Repeatedly, Herrnstein and Murray and their allies defend the tests by saying: Look at how well these tests *predict* grades, or economic success, or job performance. (That claim is, however, exaggerated.)[17] The problem is that we do not know whether it is intelligence that is predicting outcomes; we only know that the tests predict them.

A second kind of validity is *criterion* validity. A test has criterion validity if it correlates with another test that we already believe is valid. Why would we seek a valid test if we already had one? The typical reply is that the newer one may be cheaper, quicker to administer, or otherwise simpler. Note that simpler tests save schools and agencies money, but the cost falls on the people who take the tests. The correlation between an expensive and a cheap test is not perfect. Thus, a more expensive test may be worth it to the person denied a position because the test that was used was second best.

The set of assumptions needed to establish criterion validity are shown in figures 2.3(a) and 2.3(b) Figure 2.3(a) shows a statistical correlation between an expensive indicator and a cheap indicator, say ten hours of personal portfolio evaluation and a three-hour SAT. Figure 2.3(b) shows that, just as in the predictive validity case, there are two assumptions for every observed correlation. The assumptions are that both the expensive indicator and the cheap indicator are caused by some unobserved third

Criterion validity also makes a causal assumption.

(a) Criterion Correlation

10–hour Portfolio
(expensive)

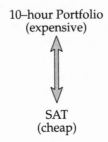

SAT
(cheap)

(b) Intelligence Interpretation

10–hour Portfolio
(expensive)

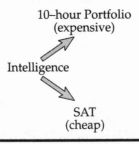

Intelligence

SAT
(cheap)

2.3. Interpreting Criterion Validity

entity, here intelligence. We might instead assume that the entity that determines both the expensive and the cheap indicator is parents' socioeconomic status or almost anything.

Under most circumstances, criterion validity and predictive validity are the same.[18] In both, assumptions outnumber information two to one. Because assumptions outnumber information two to one, there is only one way for someone to be confident of their validity—one must make a leap of faith. But faith is not the only answer to the validity dilemma; there is a third kind of validity: *construct* validity. Construct validity exists when a test measures the concept it is supposed to measure. To establish that, the test maker must specify that content ahead of time. This can be assessed not simply with statistics, but by substantive analysis. The inability to evaluate construct validity by statistics has rendered that validity invisible to many psychometricians.

Psychometricians, however, commonly argue that there is no necessary relation between the manifest content of a measure and the phenomenon that is being measured. For example, Jensen contends that the falling column of mercury in a barometer does not remotely resemble the phenomenon that it predicts, rain.[19] We need not therefore expect that an intelligence test look anything like real-world intelligence. We agree that resemblance is not necessary; but connection *is* necessary. For example, the connection between a falling column of mercury and rain is physical. A falling barometer reflects falling local air pressure, which, given the higher air pressure in surrounding locations, creates a wind pattern that leads to a lifting of air. As the surface air rises it cools and, if there is enough moisture, leads to rain. Thus, the barometer does not predict rain because it looks like rain but because there is a well-understood causal connection between the barometer and on-going air patterns.[20] Hence, it is not sufficient for the researcher simply to assert that there is a connection between the measure, questions on a test, and the construct, intelligence. If the only evidence that something real has been measured is predictive validity, then the claim is weak. By attending to construct validity, the case for the connection can be made explicitly. The psychometric paradigm, however, does not attend to construct validity; it really does not address whether what it measures *is* intelligence. (Below and in chapter 3 we do address that issue.)

A related problem is the inattention psychometricians have given to the "population" of intelligence-demanding tasks. A fundamental notion in survey research statistics is that, before one draws a sample of people, one must be able to specify the population from which the sample is drawn. Otherwise, one cannot be sure what the sample represents, and then every inference one makes from the sample is, in principle, indefensible. The analog in psychological testing is that one should specify the population of tasks to be tested before one constructs the test, that is, draws a sample of test questions. Testers have known this for a long time, but many psychometricians still contend that one need not specify the population of skills before making an intelligence test. Their argument is akin to saying one need not identify the nation under study before conducting an electoral poll. Many psychologists now reject such faulty reasoning. As we shall see below, many have moved beyond the limits imposed by psychometric assumptions explicitly to study "everyday" tasks as measures of intelligence.

When the psychometric paradigm first began to take shape, neglecting construct validity was probably wise. Researchers were very confused about how to define, measure, or shape intelligence. But now the inability of psychometricians to say more than "intelligence is what intelligence tests measure" suggests that construct validity is still not relevant to the mainstream of psychometric work. In the psychometric paradigm, intelligence is a black box, visible only in test results. No one can specify what is in that box (although some psychometricians provide after-the-fact interpretations of what it is that their tests measure, such as an ability to "figure out").[21] Therefore, no one can say what policies the research implies. One cannot embrace the successes of psychometrics—the construction of tests that predict well—without accepting the limitations of psychometrics— ignorance about whether and how well the tests really measure "intelligence." In terms of social policy, the limitations are key: Without knowing what the tests measure, policymakers are flying blind. Herrnstein and Murray try to have it both ways. They assert that the tests measure g as people's capacity for inferring relationships. But there is no consensus among psychometricians on the content of g.[22] The only way to be true to the psychometric approach is to remain agnostic on what the tests measure. American institutions continue to use such tests, but it is unclear whether heavy reliance on them is wise. It is much more unclear whether what they measure is intelligence.

What Does the Test Measure?

The problem of letting intelligence be whatever intelligence tests measure is demonstrated by *The Bell Curve* itself. Herrnstein and Murray claim that the AFQT—the test that is the basis of their statistical work—is an excellent measure of intelligence. Indeed, "the AFQT qualifies not just as an IQ test," they write, "but [as] one of the better ones psychometrically."[23] A group of psychometricians who defended *The Bell Curve* in the *Wall Street Journal* wrote that "while there are different types of intelligence tests, they all measure the same intelligence," implying that the AFQT serves that purpose, too.[24] But we shall see that the AFQT is much more a school achievement test than an intelligence test. It is a measure of how well test takers learned and displayed their knowledge of school subjects—and a measure as well of their interest, cooperativeness, anxiety, and experience in taking tests. By extension, then, other such psychometric tests are much the same.

Readers will find no examples of actual AFQT questions in *The Bell Curve*, but we will provide a few here. The test consists of four subsections, amounting to 105 questions. (As we noted earlier, the 105 questions Herrnstein and Murray used were part of a longer, 333-question, three-and-a-half hour battery.) The four AFQT sections are:

1. Section 2, Arithmetic Reasoning, composed of word problems using arithmetic skills—30 questions with a 36-minute time limit

2. Section 3, Word Knowledge, composed of 35 vocabulary words in 11 minutes

3. Section 4, Paragraph Comprehension, composed of questions referring to short paragraphs—15 items in 13 minutes

4. Section 8, Mathematics Knowledge, composed of questions testing algebra and higher mathematical skills—25 items in 24 minutes

Four examples illustrate the questions. (These are simulated versions of what remain confidential questions.) Each of these examples is rated at about average difficulty. Recall that the target population is high schoolers.[25]

1. *Arithmetic Reasoning*: If a cubic foot of water weighs 55 lbs., how much weight will a 75½-cubic-foot tank trailer be carrying when fully loaded with water?
   (a) 1,373 lbs
   (b) 3,855 lbs
   (c) 4,152.5 lbs
   (d) 2,231.5 lbs

2. *Word Knowledge*: "Solitary" most nearly means
   (a) sunny
   (b) being alone
   (c) playing games
   (d) soulful

3. *Paragraph Comprehension*: People in danger of falling for ads promoting land in resort areas for as little as $3,000 or $4,000 per acre should remember the maxim: You get what you pay for. Pure pleasure should be the ultimate purpose in buying resort property. If it is enjoyed for its own sake, it was a good buy. But if it was purchased only in the hope that land might someday be worth far more, it is foolishness.

Land investment is being touted as an alternative to the stock market. Real estate dealers around the country report that rich clients are putting their money in land instead of stocks. Even the less wealthy are showing an interest in real estate. But dealers caution that it's a "hit or miss" proposition with no guaranteed appreciation. The big investment could turn out to be just so much expensive desert wilderness.

The author of this passage can best be described as
(a) convinced
(b) dedicated
(c) skeptical
(d) believing

4. *Math Knowledge*: In the drawing below, *JK* is the median of the trapezoid. All of the following are true EXCEPT

(a) $LJ = JN$
(b) $a = b$
(c) $JL = KM$
(d) $a \neq c$

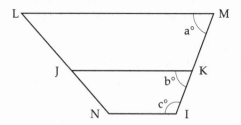

On face value, these questions do not measure test takers' intelligence, their "deeper capability . . . for 'catching on.'"[26] Mostly, they measure test takers' exposure to curricula in demanding math and English classes. They remind us of pop quizzes in high school. Two scholars, Darrel Bock and Elsie Moore, who wrote the authoritative book on this administration of the AFQT, describe the section on paragraph comprehension as "lean[ing] rather heavily on general knowledge. A well-informed person has a good chance of answering many of the items correctly without reading the paragraph. This means that the better educated . . . , having both the benefit of reading the passages and already knowing many of the facts contained in them, should have found this test very easy." (The importance of general knowledge for scoring well on IQ tests goes back a long time. In the original army "alpha" tests used in World War I, 10 percent of the questions required familiarity with national advertising campaigns, as in: " 'There's a reason' is an 'ad' for a: drink, revolver, flour, or cleaner?")[27] Bock and Moore say of the section on math knowledge that the answers would be "known only to persons who had some exposure to high school algebra and geometry, or who had studied textbooks on these subjects."[28] This seems a far cry from measuring intelligence as the "ability to learn from experience."[29] After looking at similar items, political scientist Andrew Hacker concluded, "At best, *The Bell Curve* authors have identified not a genetic meritocracy, but what might be called a testocracy: individuals possessed of a specialized skill which, on further examination, has little or no relevance to most human endeavors."[30]

It is not necessary to use school curricula questions for measuring "intelligence." For example, social psychologists Melvin Kohn and Carmi Schooler have measured what they call "intellectual flexibility" with items like: "Suppose you wanted to open a hamburger stand and there were two locations available. What questions would you consider in deciding which of two locations offers a better business opportunity?"[31] Other scholars have also realized that the kinds of questions used in the AFQT and in most other similar aptitude tests totally miss what they call "everyday" or "practical" intelligence, the kinds of ability that mature people develop, long after high school, as they deal with the complexities of real life. These researchers have also formulated other ways of testing people's abilities to figure out solutions to problems, ways not wedded to school curricula.[32] Such tests may be less finely differentiating, more qualitative, and more labor-intensive than psychometric tests such as the AFQT—and their results are different—but they also would seem to be more valid measures of intelligence broadly understood.

Researchers have also found that such paper-and-pencil tests as the AFQT are limited in predicting how people apply their knowledge in practical situations. *Context* matters. Children, for example, given an abstract logical problem to solve have great trouble; embed that same problem in a video game and they often do brilliantly. In another example, child street vendors in Brazil have trouble solving abstract math problems, but they do fine when those same problems are presented as commercial transactions. Middle-class Brazilian children react in the opposite way; they do better with the task phrased as a school test. Perhaps most discouraging to us as university faculty are findings that students who do well in classes on research design do not do well when presented with analogous logical problems outside the classroom. The point is that there is not much transfer between academic intelligence and everyday intelligence.[33]

Given the limits of psychometrics, why did Herrnstein and Murray not choose another perspective on intelligence? The psychometric paradigm is not only scientifically limited, it is also limiting because it does not provide the resources needed to make policy prescriptions to improve intelligence. It is possible, as we shall see in the next section, to identify policy prescriptions using other approaches to intelligence. Had Herrnstein and Murray adopted an alternative perspective—information processing, for example—that does allow one to speak about action, the title of their book and probably much else in it would have changed.

Psychometric reasoning has become both circular and self-reinforcing. Psychometricians found that different tests intercorrelated and from that hypothesized that they all reflected a common factor, g. They refined the tests, making this statistical component, g, more and more distinct, stable, and sharply differentiating in the tests. Educational institutions and employers increasingly used highly "g-loaded" tests for admission and placement decisions. Precisely for that reason, the *predictive* validity of such tests has probably increased over the years. (Yet SAT scores, for example, do not predict college grades any better than high school grades do.)[34] The more institutions sort people by test scores, the better the test scores predict sorting. This predictive validity is then taken as a sign that the tests must be measuring intelligence and legitimates further refinements in the tests, that is, making the tests even more narrow, more stable, and more differentiating. All this occurs, however, without any clear evidence that the tests actually measure what they purport to measure. And because the tests are refined to magnify differences, there is little evidence that the original differences are as large as they now seem to be. Finally, even all of this would not be of such concern were psychometrics put into perspective. Who would care that the tests are designed to make minute differences appear mountainous if the discussion and use of the tests kept their limits firmly in the foreground?

Most college applicants and administrators probably ignore the ambiguity of the tests and the mountains-out-of-molehills strategy of test construction as they open their mail from Educational Testing Service. But test construction in the psychometric paradigm magnifies differences among people and so impedes our understanding of inequality. From inside the psychometric tradition the reasoning may appear impeccable: Specific, stable, and highly differentiating tests predict outcomes, so they must be measuring something real. No wonder, then, that psychometrics may lead one to believe in the fatedness of life. Trapped in that circle of reasoning, anyone might lose sight of other possibilities and become convinced of the hopelessness of ever ameliorating inequality. Fortunately, there are other ways, ways that identify intelligence and are more hopeful about changing it.

The Political Arithmetic of Testing

Tests appeal to decision makers and the public alike because they appear to be objective. The SAT is a prime example in many respects. It is said to be heavily "g-loaded," predicts college grades modestly,[35] and appears to be neutral. It can be useful when applied properly, but too often the test is abused. An important case in point is the recurring controversy about the relative SAT scores of whites and members of minority groups at some universities. Many people who care about college admissions point out that minority freshmen have average SAT scores far below those of white freshmen. The gaps that they point to are often quite large.[36] The minorities, Herrnstein and Murray argue, must have gotten an edge in the admissions process. Otherwise, they ask, how could freshmen with scores so far below those of the average white student get into this university? But there is a statistical fallacy here. It is impossible to tell how much of an edge minority students have been given simply by looking at the difference in average test scores. This is so even if test scores were the only criterion for admission. The groups will differ in scores among freshmen, absent any racial preference at all, because minority applicants score lower on average than do whites. (*Why* minority applicants have lower average scores is the subject of chapter 8.)

As a simple example, imagine a very small college that will admit ten freshmen. The college receives twenty applications from eighteen white high school graduates and two black ones (roughly the proportions nationwide). The twenty students have the following SAT scores:

1300	1225	1200	1175	1150	1125	1100	1075	1050*	1025
1000	975	950	925	900	875	850	825	800*	750

The two black students are identified with asterisks. The average of these twenty SAT scores is 1014 (just 14 above the national norm).[37] The average for the eighteen white students is 1018; the average for the two black students is 925. The 93-point difference is 22 points smaller than the national difference. (These twenty SAT scores are surprisingly realistic considering that we are working with such a small number of "observations.")

If the dean of admissions at this little college uses nothing but test scores to admit students, then the ten students in the top row will be the ones who are admitted. The process, if followed, would be completely race-neutral, just as the critics of current practices would like it to be. The two black students were given no edge. Yet when we examine the entering class, we see that the mean for the nine white freshmen is 1153, and the black freshman's score is 1050—leaving a gap of 103 points. The race-neutral admission process simply passed on the preexisting difference in test scores from the applicant pool to the freshman class. Any race-neutral process will.

We have repeated the exercise with much larger and more realistic numbers.[38] The conclusion is the same. Race-neutral admission practices that rely solely on SAT scores would yield an entering class that shows a disparity in scores. Race-neutral selection processes pass disparities in the applicant pool through to the freshman class. Therefore, we cannot read a gap in test scores as if it reflected an edge that the admission process gives to some students at the expense of others. In part, it reflects the disadvantages that suppress the measured achievement of some groups, especially blacks and Latinos. It might also reflect the tendency of minority applicants to score higher on other criteria of admission, such as grades and class rank.

Suppose the system does give some edge to minority applicants through an affirmative-action system? Is that edge simply added on top of the "naturally occurring" difference that we just saw race-neutral admissions pass through? No.[39] Although the SAT gap between blacks and whites in the freshman class would be larger under affirmative action than it would be under race-neutral admissions, the SAT gap would not increase by the amount added to black students' scores.

Similarly, the fact that the average test score among freshmen of Asian American descent is higher than that among white students does not prove that universities are discriminating against Asian Americans. It, too, reflects the distribution of test scores in the applicant pool. The admission process may simply reflect the higher average scores that Asian American applicants bring to the freshman class.

In short, we cannot actually tell much about the race-sensitivity of the admission process from the racial disparity of the outcome. A race-neutral admission process does not result in equal means for students from all groups, even if test scores are the only criteria for admission (most schools use other criteria as well). The only conclusion that holds in general is that the profile of SAT scores in the freshman class reflects the profile of SAT scores in the applicant pool.

We draw two lessons from this exploration of the political arithmetic of testing. The first is the substantive point that the discussion of racial and ethnic preferences in college admission is based on a statistical fallacy. The average score of freshmen from one group might be lower than the average for freshman from some other group for a variety of reasons; it does not imply that applicants from the lower-scoring group were given preferential treatment. Our second lesson is that the apparent objectivity of tests invites their abuse. The claim that affirmative action is reverse discrimination would be much harder to make if those who make it could not cite group differences in test scores. Ironically, affirmative action's critics make themselves seem authoritative and objective in the act of committing a fundamental error. In fact they are neither authoritative nor objective. Their use of the numbers is selective and calculated to persuade, not inform.

THE INFORMATION-PROCESSING VIEW OF INTELLIGENCE

Despite the inclination of Herrnstein and Murray and much of the media to make psychometrics appear to be the only perspective on human cognition, it is not. There are several alternative approaches, ones that posit multiple intelligences or that stress the relativism of intelligence, for example. To illustrate the alternatives, we look closely at one: information processing.

In contrast to the psychometric school, scholars within the information-processing tradition begin with a definition of intelligence. A full treatment of this paradigm is beyond our scope, but a brief discussion should suggest some of the ways in which the information-processing school breaks through the psychometric circle. The psychometric and information-processing traditions have very different implications for how we think about inequality and what we might do about it.

An Information-Processing Definition of Intelligence

One noted information-processing adherent, Robert Sternberg, defines intelligence as "purposive adaptation to and selection and shaping of real-world environments relevant to one's life. Stated simply, it is mental self-management."[40] Note that this definition does not dismiss psychometric testing. Depending upon the test takers' familiarity with tests, intelligence tests may tell us something about how proficient test takers are at adapting to an environment. The test itself *is* an environment, although a particularly sterile one. Even at best, however, standard psychometric tests tell us nothing about how proficient test takers are at *shaping* or *selecting* their environments. At worst, intelligence tests tell us that the test taker who shapes the test environment (i.e., interprets the questions differently) or who refuses to spend any time on the questions (i.e., selects another environment) is a failure. Thus, a psychometric test may tell us some things but cannot tell us other important things about the test taker's intelligence.

If intelligence is defined as mental self-management, then intelligence can be taught. The very existence of business schools attests to our confidence that management skills can be taught. To teach intelligence, researchers must discover what mental processes are invoked when people solve problems. They must learn how people manage their mental resources. As these processes and strategies become understood, researchers can devise ways to test people's use of them and to teach people to use them better. An example will suggest what may be learned from such research and how that learning can influence intelligence testing and training.

Many psychometric intelligence tests contain analogies. On their face, these types of questions should directly test what Herrnstein and Murray have defined as intelligence, the "general capacity for inferring and applying relationships" (p. 4). Sternberg argues, however, that *how* people reach the inference is also important; two people who arrive at the same answer may have used different processes to get there. Consider the following linear syllogism; it is simpler than an analogy, but still requires that the test taker grasp a relationship:

Mark is taller than Adam.
Adam is taller than Jerry.
Who is shortest?

One can use either a verbal strategy or a spatial strategy to answer this question. In the verbal strategy, the test taker attends to the literal relationship between the objects. In the spatial strategy, the test taker might visualize or draw a map of the relationships. But this question is unlikely to appear on a test of spatial intelligence. However, a question with a similar logic might appear on a spatial test. Consider the following spatial analogy (which answer, a to d, belongs in part 4?):

Part 1 $\quad|\quad|\qquad$ Part 2 $\quad|--|$
Part 3 $\quad$ 0 $\quad$ 0 $\qquad$ Part 4 $\quad$ __?__

(a) 0 $\quad$ 0
(b) 0--0
(c) ● $\quad$ ●
(d) ■--■

The question requires one to infer from the relationship between part 1 and part 2 to the relationship between part 3 and part 4. This is an analogy problem presented in spatial terms. Many test takers answer the question in spatial terms, but some translate it into a verbal problem and then solve it. An information-processing analyst would point out that three distinguishable skills are being tested in these examples: (1) the ability to recognize relationships; (2) the ability to apply such relationships to another domain; and (3) the ability to translate a given problem into the terms that are personally easier to solve.

Two key inferences may be drawn from the analogy examples: First, intelligence *can* be taught. Success in such tests depends in part on recognizing relationships. And something can be recognized only if it has been seen before—that is, if one has been *educated* about such relationships. Sternberg lists thirteen different kinds of relationships that can be found on one common test, the Miller Analogies Test. If these relationships can be so identified, then students can certainly be taught them. We regularly teach small children to recognize thousands of relationships among a small set of signs, the twenty-six letters of the English language. (Imagine how difficult it would be to read or write had one been exposed to only half of the letters.) Second, in order to select a strategy that maximizes one's chance of answering correctly, one must know one's own strengths and weaknesses. Some may find it helpful to translate a problem into another format before proceeding. Good mental self-management can identify one's strengths and weaknesses. Also, weaknesses can be strengthened, just as strengths can atrophy, so that training can shape success on such intellectual tasks.

In sum, intelligence in the information-processing framework is mental self-management, and mental self-management involves selecting, adapting to, and shaping real-world environments. These intelligence skills can be taught and trained; they are neither fixed nor singular as the psychometric view assumes. The implication for the entire debate on inequality is clear: To the extent that intelligence does affect who gets ahead, so does teaching.

Measuring Intelligence

Critics often complain that psychometric intelligence tests are narrow. Our earlier discussion should suggest why that criticism does not faze psychometricians: the tests are supposed to be narrow. But good measures of intelligence must take into account the many aspects of intelligence that are neglected, deemphasized, or even negatively evaluated in psychometric tests (e.g., creativity). We can see the multiplicity of intelligence in the existence of "idiots savants" who can perform brilliantly in some ways and are mentally deficient in others. We can also see it in recent brain research that points to many different centers for information processing rather than a single central processor. An example of a dimension of intelligence unexplored in psychometrics but critical to information processing is defining a problem. For example, defining a problem is probably a task that requires intelligence. In normal life, it is the rare problem that arrives neatly, with explicit questions and "the" four alternatives placed just below each question, one of which is surely correct. A good intelligence test should test the ability to define a problem.

Good tests in the information-processing tradition also test the intellectual *process*. Not content to leave intelligence as a black box, information-processing researchers study how people process information. One line of research that may reveal much about how people process information, as well as shed light on how to improve human performance, is the comparative study of experts and novices.

Experts know more facts about a subject than novices do, but that distinction does not fully explain the differences in performance between experts and novices. The two apparently process information and think about a problem in qualitatively different ways. Yet because every expert was once a novice, somehow novices must *learn* to think like experts. Differences in how novices and experts process information could therefore be taught, as a part of mental self-management.

In one well-designed study, Jan Maarten Schraagen investigated how novices and experts differ in their use of problem-solving strategies.[41] Schraagen studied four groups: beginners, intermediates, "design experts," and "domain experts." The test was to design an experiment to answer a specific research question. The domain experts were experts in the specific research topics; the design experts were expert in how to design experiments. (Thus, Schraagen was able to determine whether knowledge of content alone explains the difference between novice and expert.) Schraagen found that both kinds of experts brought to bear problem-solving strategies that beginners and intermediates did not. Although domain experts were more proficient than design experts, owing to their vast store of knowledge about the domain, design experts resembled the domain experts in *how* they reasoned. In short, there were qualitative differences in the structure of reasoning between experts and nonexperts.

For example, experts tended to use a procedure known as *chunking*—linking information so that remembering one element in a chunk calls forth the full chunk of information. The opposite of chunking is to remember each piece of information as an isolated element. By chunking information, experts not only sped up retrieval but also assured that retrieval of any one item in a chunk would call forth other needed items. Another example comes from chess: A. D. de Groot compared how quickly master players and amateurs could recall the locations of chess pieces on a board. The masters performed much better than the amateurs. But when the pieces were placed not in a recognizable chess-game pattern but randomly on the board, there was no difference between the two groups. Why? Masters did better than amateurs not because their brains were inherently superior but because they had learned expectable patterns that emerge in a game (e.g., the positions of a few key pieces will trigger the memory of which kind of defense is being used). Remembering one piece of the pattern stimulates recall of the entire pattern. When the pieces are randomly placed, learning matters much less.[42]

As these and other expert tactics are studied further, researchers are learning how to teach cognitive strategies explicitly. For example, Jonathan Baron found that children labeled as retarded could perform almost as well as "normal" children on memory tasks after they had been trained to use the same kinds of memorizing and monitoring strategies other children use. Lest one think these strategies can only aid students at the lower range, explicitly teaching monitoring strategies (i.e., mental self-management) also increases the success of college-level math students.[43] We teach such strategies whenever we teach novices to become experts.

There are many more aspects to the information-processing perspective on intelligence testing than we can treat here.[44] The elements we have identified show how aspects of intelligence are beyond the scope of psychometric tests but flow easily from the information-processing tradition. Psychometric tests are limited because they tap only a small part of intelligence and then magnify small differences. Moreover, the information-processing paradigm does not discard the findings of the psychometricians but puts them into perspective and then adds additional insights. This is the familiar process by which one paradigm supplants another.

Implications

Particular policy prescriptions that psychometricians might dismiss make a lot of sense from an information-processing tradition, among other alternative paradigms. Herrnstein and Murray dismiss educational intervention as a waste because they believe that how intelligent people become is largely fixed at birth or shortly thereafter. Thus, *The Bell Curve*'s only serious educational proposal is to invest more in "gifted" students, not "dull" ones. But because research shows that intelligence can be taught and learned, it is sensible to invest resources—time, talent, attention, and money—into educating all our people. We can improve Americans' cognitive skills even more than we already have (see next box). We can teach intellectual self-management in school, tailoring instruction to each student. The information-processing tradition, in contrast to the psychometric tradition, recognizes that people can grow intellectually throughout their lives, even long after their formal education has ended. In chapter 7, we review several specific, concrete ways in which our social policies shape Americans' intelligence.

The Brilliance of Americans Today: A Puzzle

Psychometricians have discovered a dramatic and encouraging but, for them, embarrassing finding: According to their tests, people around the world are far smarter now than were their parents and grandparents. Little in the psychometricians' genetic paradigm allows them comfortably to explain this change.

The "Flynn Effect," named after James Flynn who brought together all the evidence, shows a rise in Western countries of roughly fifteen IQ points in one generation. Another scholar estimates that Americans of the 1970s were twenty-two points smarter than Americans of the 1890s. This is an amazing change, implying that the average American of about 1895 would be considered "dull" in 1975 and that a large proportion of Americans today would have been near-geniuses in their great-grandparents' day.

Psychometricians who hold that intelligence is genetically determined have great trouble accepting this happy conclusion, however. Genetic change cannot account for such a rapid increase, no matter how optimal breeding patterns may be. (Most of the sources Herrnstein and Murray rely on claim that breeding patterns actually have been driving intelligence downward. But the historical data are clear and imply that intelligence has been sharply elevated by changing social environments.)

Psychometricians have taken different routes to escape this fundamental contradiction. Flynn argues that the IQ tests are faulty; they do not really measure intelligence but something loosely related to intelligence, "abstract problem-solving ability." Herrnstein and Murray tentatively suggest that the change is partly real (but temporary) and partly the result of people having learned how to take IQ tests. Miles Storfer accepts that the change is real and concludes that evolutionary theory must therefore be wrong: Lamarckism is possible and people can pass on learned skills biologically!*

If "intelligence is what intelligence tests measure," but the results seem implausible, what can a psychometrician do? These are the sorts of contradictions that unsettle and eventually break down a paradigm.

* Flynn, "Massive IQ Gains"; Herrnstein and Murray, *The Bell Curve*, pp. 307–9, passim; and Storfer, *Intelligence and Giftedness*, chaps. 5, 18. There is also evidence that social classes are converging in intelligence—another contradiction to Herrnstein and Murray (Weakliem et al., "Toward Meritocracy?").

CONCLUDING REFLECTIONS

Herrnstein and Murray discuss information processing and other alternatives to psychometrics in their introductory chapter, but by the end of it they claim to have good reason for using only the psychometric tradition. They argue that their focus is on the "relationship of human abilities to public policy" (p. 19). In doing so, they deny an interest in the *development* of human abilities. Indeed, Herrnstein and Murray admit that, had they an interest in cognitive development, they would have spent more time using the information-processing tradition (p. 20). So we agree with them on this point: The information-processing approach is far more applicable than is psychometrics when cognitive development is the issue. To the extent that Herrnstein and Murray address the (im)mutability of intelligence, however, *cognitive development is precisely the issue* and the psychometric tradition is not appropriate.

With the exception of criticizing affirmative action and supporting programs for the "gifted," *The Bell Curve* gives little attention to how the educational system might be altered to improve cognitive skills. A study of intelligence is remiss in neglecting the inner workings of the only institution in this country whose very raison d'être is improving people's abilities to solve intellectual and life problems. The only justification for ignoring schooling appears to be that Herrnstein and Murray believe intelligence to be immutable. Yet they use a theory of intelligence that they themselves admit is of limited value in assessing mutability.

We, the authors of this book, are in a bind. We have argued that the way intelligence is analyzed in *The Bell Curve* is wrong. Intelligence is not a single, essentially immutable entity. That should be sufficient to dismiss the rest of *The Bell Curve*. For some psychologists, it is. One wrote that the "most benign interpretation" of all the errors in that book "is that the authors were simply operating with outmoded psychological notions . . . an old-fashioned psychometrics and almost equally outdated behavior genetics."[45] But to respond fully to Herrnstein and Murray's arguments, we will in the next two chapters set aside these reservations and accept their psychometric stance. We will look closely at exactly how they applied psychometrics to the topic of inequality.

When Americans discuss the connection between intelligence and social policy, they need not be bound by the limits of the outdated psychometric paradigm. They can transcend the fear-laden politics that seem to accompany the zero-sum result of psychometric reasoning. Newer understandings of human psychology offer better ways of understanding the psychological factors that advantage some people over others in the race for success. They also offer a realistic hope of building a smarter and better society.

But Is It Intelligence?

Many American parents have had the experience of trying to help a teenager with his or her homework and discovering that they have forgotten how to calculate the volume of a sphere or could not recall "lowest common factors." Even subjects like grammar seem embarrassingly vague to them, at age forty or so. Yet the teenager often whizzes through the same problems. According to psychometric tests, including the one used in *The Bell Curve*, the teenager is more intelligent than the parents. This is the same teenager who is always late getting off to school, cannot figure out how to get from point A to point B without being driven, is socially awkward, and cannot remember to turn out the lights in his or her own room. How can the teenager be more intelligent? Only by defining intelligence as doing well on tests that measure learning in school and that do not measure doing well on other life tasks. It is this limited notion of intelligence that forms the evidence at the foundation of *The Bell Curve*.

The proposition that natural inequality in intelligence explains economic inequality among people is based on the assumption, which the previous chapter demonstrated to be dubious, that there is a single, fixed trait of "intelligence." If one nevertheless proceeds on that assumption, the researcher must then find an instrument that accurately measures such intelligence in order to test the proposition. In this chapter, we will examine the instrument used in *The Bell Curve*, the AFQT. We will show that the AFQT is a poor measure of innate intelligence and instead reflects the social environment that shapes people's academic performance, largely their schooling. Because schooling helps people get ahead, this is confirmation that the social environment matters most in explaining inequality. And the further implication is that we can, through training, raise people's performance on the sorts of tests that are admission tickets to upward mobility in our society.

The reader will recall that the central evidence in *The Bell Curve* comes from the National Longitudinal Survey of Youth's repeated interviews with over 12,000 young respondents. In 1980, when these respondents were between the ages of fifteen and twenty-three, they took the AFQT.

Richard Arum coauthored this chapter.

Herrnstein and Murray use scores on this test as their measure—as their *operational definition*—of intelligence. Through a few hundred pages of *The Bell Curve* they argue that low intelligence, as indicated by low scores on the AFQT, determined which NLSY youth dropped out of school, became poor, became criminals, and otherwise had problematic young adulthoods.

The AFQT, however, was not designed to measure intelligence or cognitive ability. It was designed to predict how well high school graduates would do in the armed forces. It was written to measure *school achievement*—more precisely, to measure how well teenagers attained high school–level math and reading skills.[1] (As many critics have noted, the original Binet tests were designed to identify children with problems in school and specifically not as a test of inherent intelligence. American psychologists turned it into a measure of "IQ.") Furthermore, the AFQT is problematic even as a gauge of a person's school learning, because test takers' scores reflect other factors as well, such as the instruction they received outside of school, their social backgrounds, and their motivations.

The Content of the AFQT

In chapter 2, we discussed how the psychometric tradition that Herrnstein and Murray follow works from the test backward to the concept of intelligence. Psychometricians did not identify g, the general factor for intelligence, by observing people behaving intelligently; they derived it from statistical analyses of test questions, from the tendency of people who answer one question accurately to answer others accurately. It is a concept built from the test upward. (Herrnstein and Murray admit this point but later forget it: "The evidence for a general factor in intelligence was pervasive but circumstantial, based on statistical analysis rather than direct observation. Its reality therefore was, and remains, arguable" [p. 3].) What *does* the AFQT measure?

In chapter 2, we looked at a few questions from the AFQT itself. They clearly tested an examinee's command of school curricula. Here are a few more examples:[2]

Two partners, X and Y, agree to divide their profits in the ratio of their investments. If X invested $3,000 and Y invested $8,000, what will be Y's share of a $22,000 profit?

(a) $8,250
(b) $16,000
(c) $6,000
(d) $5,864

Reveal most nearly means
(a) cover again
(b) turn over
(c) take away
(d) open to view

The greatest common factor of 16, 24, and 96 is
(a) 8
(b) 2
(c) 16
(d) 12

If j and k are positive whole numbers and $j + k = 12$, what is the greatest possible value of jk?
(a) 6
(b) 36
(c) 32
(d) 11

As before, we see that the AFQT questions are manifestly about *school* tasks.

Still, intelligence as most people generally understand it probably contributes to doing well on the AFQT. Youths who process information better probably learn more in school and so test better—as do youths who sit still, pay attention, and care more. At least two problems, however, cloud even this rationale for the AFQT as a test of intelligence. First, school subjects comprise but one of many realms of life in which intelligence might matter. This test ignores those other realms—social relations, business, mechanical arts, and so on—and thus is, at best, a partial test. Second, for youths to display intelligence on this test requires having been at least exposed to the school subject matter. If intelligent youths have not been instructed in the material—and it matters not whether the deficiency arises because of lousy schools, lousy homes, lousy attitudes, or simply being too young to have taken these subjects yet—then they will not score well. (See, for example, the case of the Founding Fathers on page 58.) Moreover, of the four subtests in the AFQT—arithmetic reasoning, word knowledge, paragraph

---

WERE THE FOUNDING FATHERS "DULL"?

In the course of our history, Americans' "numeracy"—command of numbers and numerical operations—has greatly increased. At the time of the Revolution, most Americans had little familiarity with, or interest in, exact calculations. The few who formally learned arithmetic struggled with unsophisticated teachers and crude textbooks that focused on mnemonic rules for converting currencies and measures.*

If the AFQT had been administered to Revolutionary-era Americans, to the Founding Fathers perhaps, most probably would have scored in the lower range. Would we then conclude that they were less "intelligent" than today's high school students? Probably not. We would conclude that today's students *know* more mathematics than did George Washington and his fellows because today's students are *taught* more mathematics than were the Founding Fathers. And that is just what the AFQT measures: what students have been taught.

* Cohen, *A Calculating People*.

---

comprehension, and mathematics knowledge—the two math components make the greatest difference in the final scores.[3] These are subjects that probably most require having enrolled in the right classes, having had good teachers, and having paid attention.

The AFQT questions themselves testify that what is being assessed is, foremost, environmental, not "natural." Vocabulary words, profits and ratios, greatest common factors—these are topics that children are taught, usually in school. The AFQT is also a test of environmental influences in the sense that it has a very specific substance, curricula taught in late-twentieth-century American high schools. Similarly, it also taps vocabulary and concepts specific to middle-class American homes in the late twentieth century.[4] In another place or time, an armed forces screening test might consist of, for example, reading signs of an animal trail, knowing how to find water, staying upwind of prey, and so on.[5] Our critique here rests on questioning the AFQT's content validity (see chapter 2) as a test of *g* by simply reading the test. There is other evidence, as well, that the AFQT is a better measure of social background than of "native" intelligence.

Schooling, Age, and the AFQT

Statistical evidence supports reading the AFQT as essentially a test of mastering school curricula. Following Herrnstein and Murray, we reanalyze the NLSY survey, focus on white respondents only, and use the test scores they calculated from the 105 AFQT questions. They call those transformed scores "zAFQT," referring to standardized scores designed to conform to a bell curve.[6] (See chapter 2 for how Herrnstein and Murray constructed a bell-curve distribution from original test scores that were not distributed in a bell curve.)

Cognitive ability, if it is like other developmental traits, should grow as people grow (at least until the debilitation of advanced age). When doctors test children on, say, hand-eye coordination, they take into account the children's ages. Similarly, psychometricians do not simply compare children of different ages on raw IQ test scores, because they presume that children score higher as they age. Psychometricians "standardize" IQ test scores for age. Age needs to be taken into account in the AFQT, too, because the NLSY gave the test to youths ranging from fifteen to twenty-three years old. Herrnstein and Murray, for the most part, dealt with age in a roundabout fashion. In their statistical analyses showing that the AFQT predicts outcomes such as poverty, they introduced the respondent's age at the time he or she took the test as a "statistical control variable." At this point, all that means for us is that Herrnstein and Murray, in effect, assumed that test takers' AFQT scores increase as the test takers age, and so they wanted to hold it constant.[7] This makes sense from the psychometric perspective: Intelligence is inborn or nearly so and matures in a developmental manner, just as eye-hand coordination does.

But if it is instruction in curricula rather than biological development that is critical, then a different attribute needs to be controlled—not test takers' ages, but how much math and vocabulary they had already been taught when they took the test. The NLSY provides researchers with only crude measures of the test takers' histories of instruction: the number of years of school the test takers had completed at the time of the test and whether or not they had been in an academic track in school. These are crude measures, in part because they do not reflect the quality of instruction the test takers had received. Graduates of Andover, the Bronx High School of Science, a weak inner-city high school, or a remedial high school for troubled children all score "12" for years of schooling completed. These measures are crude also because they do not reflect students' exposure to informal instruction at home, during the summer, in travels, and so on. Children who are read to at home or whose parents encourage them to pay the bill in a store will be ahead of the curve in vocabulary and math. Yet it turns out that even these two crude measures of instruction—years of schooling and having been in an academic track—correlate much more closely with test takers' AFQT scores than do their ages.

(Readers will recall that a positive correlation coefficient, between 0 and 1, means that people who are high on one trait tend to be high on the other. A negative correlation, from 0 to −1, means that people high on one trait tend to be low on the other. The higher the absolute value of the correlation, the stronger the connection. Practically speaking, correlations, which are symbolized as r, above $r = .20$ to .30—or below $r = −.20$ to −.30—suggest that there is a noteworthy correspondence between how people scored on one measure and how they scored on another.)

The table below shows how test takers' AFQT scores correlate with their ages, on the one hand, and with two indicators of instruction, on the other. Clearly, one can predict how high someone scored on the AFQT pretty well by knowing how far the person had gone in school and what track he or she had been in; one can hardly predict AFQT score at all by knowing the test taker's age.

Correlations of AFQT score with. . . .

test taker's age at the time the test was taken	.16
years of education completed at time test was taken	.54
whether the test taker had been in an academic track	.45

Scores on the AFQT strongly correlate with how many years of schooling test takers had had when they took the test (.54). Another way to see that is to look at who were among the highest and the lowest scorers. The high scorers, the top 5 percent (Herrnstein and Murray's "cognitive elite"), were those who answered 101 or more of the 105 questions correctly. Over two-thirds of these high scorers had had at least one year of education beyond high school when they took the test. Of those who scored in the bottom 5 percent (the "very dull"), almost half had already dropped out of high school before taking the test.

Age, on the other hand, weakly correlates with AFQT score (.16). In fact, age correlates *negatively* with AFQT if one holds years of education constant.[8] That is to say, among NLSY respondents with grossly similar exposure to instruction, the older ones scored below the younger ones. So, for example, older respondents with twelve years of schooling scored below younger respondents with twelve years of schooling. The logical explanation for this result is that the older respondents had been out of school longer and had had more time to forget what they had learned there.[9] It makes sense because the AFQT measures instruction. If we were instead to accept Herrnstein and Murray's premise that the AFQT measures native intelligence, these results would imply that as people aged from teenage-hood into young adulthood they became stupider. Only teenagers might buy that explanation.

Herrnstein and Murray are not much concerned with test takers' ages because they assume that intelligence is immutable—fixed for all intents and purposes even before age fifteen. Herrnstein and Murray wrote, "The AFQT test scores for the NLSY sample were obtained when the subjects were 15 to 23 years of age, and the IQ scores were already as deeply rooted a fact about them as their height" (p. 130). That is a bold but untenable statement.

Think about the heights of teenagers. Take a group of fifteen to twenty-three-year-olds and line them up from shortest to tallest. Gather them together again in eight years and line them up again. In general, the order of people from shortest to tallest will have changed some, but their actual heights will have changed a lot. We cannot validly compare the heights of fifteen- and twenty-three-year-olds on any one day. The fifteen-year-olds still have growing to do, but nearly all of the twenty-three-year-olds will have reached their full height.

Now think about the AFQT scores of fifteen- to twenty-three-year-olds. These comparisons are even more problematic—as the correlation of a mere .16 shows. It implies that there is little connection between age and score. The correlation is weak because youths differ greatly in how much education they get between age fifteen and age twenty-three. People leave school all along the way during those years. If AFQT score was determined by maturation—as height is—then the correlation between age and AFQT would be as high or higher than the correlation between education and AFQT score. It is far lower, however. Those who receive more or better schooling between age fifteen and age twenty-three pass ahead of those who receive less or worse schooling. Consider two young women in the sample, one fifteen and the other twenty-three, and suppose that the fifteen-year-old has exactly the AFQT score the twenty-three-year-old would have gotten if she had been tested at fifteen. Suppose also that the fifteen-year-old will ultimately attend the identical schools for exactly as long as the twenty-three-year-old already has done. Can we expect these two women to obtain the same AFQT score on the same day? (If intelligence were fixed by age fifteen and if the AFQT measured this intelligence, then the answer would be yes.) The answer is, of course, no. The twenty-three-year-old has already finished high school and gone on to college; the fifteen-year-old is still in the ninth or tenth grade. The twenty-three-year-old can bring to the test all of the instruction she received since age fifteen, but the fifteen-year-old still has that instruction ahead of her. The apparent paradox about aging and intelligence—the comparison between teenager and parents with which this chapter opened—underlines the fallacy of taking tests like the AFQT to be measures of intelligence broadly understood.

The issue of aging and testing does not merely reveal a technical problem in *The Bell Curve*'s use of the AFQT. It speaks to the essence of what we mean by "intelligence" and to what we can do about it. Until the 1960s and 1970s, psychometricians had convinced most specialists of aging that from about age twenty onward, people became less and less intelligent, because researchers had found that older people scored considerably worse on the psychometric tests than younger people did. But those tests, like the AFQT, largely tapped school and school-like learning. In recent years, psychologists have realized that such academic tests measure only one special kind of cognitive ability; the tests ignore other psychological capacities, such as practical skills, social acuity, and wisdom. New research uses tests that measure a wider variety of "everyday" problem solving, for example, how a person would advise a friend who had a financial or medical problem. The research shows that people sustain or even expand their mental capacities well into old age.[10] The inability of the psychometric paradigm to handle this most elementary point, that in many ways mature adults are "smarter" than teenagers, casts doubt on that entire enterprise.

The age pattern in the NLSY data shows how the AFQT is really a test of schooling. So does the high correlation between AFQT score and having been in an academic track (.45). Classes in academic tracks teach more about the topics covered in the AFQT than do classes in general or vocational tracks. This should not matter if the main determinant of AFQT performance is some kind of raw intelligence. It should matter a great deal if the main determinant of AFQT score is instruction in academic material. The latter is what the evidence shows.

In some ways, the AFQT might be a good measure of instruction, but not one of native intelligence. What it captures best is how much instruction people encountered and absorbed. It does that better than does the conventional "years of education" measure, because the AFQT seems to assess educational quality and informal instruction as well as simply time in school. It taps the difference between those who spent time in classes with rich curricula, energetic teachers, motivated students, and plentiful resources and those who spent time in classes without those qualities. It taps the difference between those who are "instructed" outside the classroom and those who are not. (See, for example, the discussion of summer vacations in chapter 7.) These differences favor rich children over poor ones. Of course, the AFQT also taps the difference between those who concentrated and remember what they were taught and those who did not. Yet even these cognitive skills can be and are taught, as we saw in the discussion of information processing in chapter 2.

One response to our argument that the AFQT is a test of instruction, especially schooling, is that we have the order reversed: AFQT scores really do measure intelligence; and, as proxies for intelligence, they cause, rather than are caused by, schooling. That is to say, smart kids score high on the test and smart kids go further in school. This controversy is not a mere technical debate; it has profound policy implications. Whether the instruction we give children molds their intelligence or their intelligence molds the instruction they receive says a lot about whether and how to invest in schooling.

To support their claim that the amount of schooling youths complete only reflects the intelligence they brought to school, the authors of *The Bell Curve* conducted a complex statistical analysis. NLSY researchers had access to scores from earlier school-administered IQ tests for about one-fifth of the respondents. Herrnstein and Murray report that the higher respondents had scored on the earlier IQ test, the higher they scored on the AFQT. How many years of school the respondents had completed in the period between the two tests contributed little to the ability to predict AFQT score; all one needed to know was the earlier test score. Herrnstein and Murray concluded that AFQT scores in 1980 were caused almost solely by earlier IQ scores, not by education—and, therefore, the AFQT measures what other IQ tests supposedly measure, intelligence (pp. 589–92).

There are, however, several problems with Herrnstein and Murray's exercise. For one, using *years* of education to measure the effect of schooling ignores *quality* of education.[11] For another, they made a couple of technical errors, which, when corrected, double their own estimate of education's effects on intelligence.[12] A more sophisticated statistical analysis of the same data by two economists showed that years of schooling affected AFQT score more than three times as much as Herrnstein and Murray estimated.[13] Furthermore, Herrnstein and Murray fail to report an important piece of evidence about the predictive validity (see chapter 2) of the AFQT. Test takers' scores correlated with the number of years of school they had finished at the time they took the test at $r = .54$. Test takers' scores correlated with the number of years of school they completed *after* taking the test at only $r = .33$. The AFQT score better "predicted" past schooling than it did future schooling. That is, the AFQT measured what test takers had already learned, not their ability for future learning.

A more fundamental problem with Herrnstein and Murray's effort to separate the AFQT score from schooling is that other IQ tests also reflect instruction. Take, for example, the test used by the NLSY to assess the children of the respondents. By the late 1980s, many of the young women the NLSY had been tracking since 1978 had themselves borne children. Interviewers gave these women's three- to six-year-old sons and daughters the Peabody Picture Vocabulary Test. In this test, an adult reads a word and then asks the child to pick out from four choices the correctly corresponding picture. This most surely tests substantive learning of vocabulary—learning that may come from talking with parents, listening to parents read books, watching educational television, and the like.[14] Typical IQ tests share the property of measuring school-like learning.

Finally, we must not forget that schools often use such tests to track students. This, too, will create correlations among scores on early IQ tests, exposure to curricula, and scores on later tests, because schools place students who score well in the early tests into enriched classes and place low scorers into remedial classes. The classes students are placed in affect how much instruction they receive and, subsequently, how well they do on future tests. (See chapter 7.)

That instruction in curricula is the cause, not the consequence, of test scores can be illustrated with another sports analogy. Suppose we wished to measure people's "lifting quotient"—call it "LQ." We ask a sample of people to lift weights and use their performance to measure their LQs. Some of these people have been working out in gyms where they strengthened themselves and received instruction in weight-lifting techniques; others have barely exercised since their physical education classes in high school. Obviously, how much time people had spent in a gym would correlate highly with how much poundage they lifted in the test. Maybe high LQ causes more gym time (perhaps people who are strong choose to work out more), but it is much likelier that the causality runs largely the other way: More gym time causes high LQ scores. Similarly, more instructional time—in and out of class—causes high IQ scores. The AFQT is a measure of instruction received, absorbed, and displayed.[15] And, as such, it reflects the *environment*—schools mostly, but also families and communities—of the test takers.

This understanding of the connection between education and test scores is not only more accurate than Herrnstein and Murray's. It also encourages us to believe that we can raise test performance through more and better teaching.

WHAT ELSE DOES THE AFQT MEASURE?

The AFQT is a better measure of instruction than it is of "natural intelligence," but it is not a perfect measure of the former either. Scoring well or poorly depends on other factors as well, most notably, on test takers' motivation to display that learning.

To see this, let us look closely at the tails of the bell curve distribution, the highest- and lowest-scoring test takers. We showed in chapter 2 that Herrnstein and Murray squeezed a bell curve out of what was not at all a bell-shaped distribution of original scores. They did that because they were convinced not only that intelligence had to be distributed in a bell curve, but also that the dramatic differences in life outcomes are found between the extremes. It is the highest 5 percent, the "cognitive elite," who contribute the most to society and the bottom 5 percent, the "very dull," who are responsible for most social problems.

Who were in the upper end, the 5 percent "cognitive elite"? As noted earlier, they were overwhelmingly people who had had some schooling beyond high school before taking the test.[16] And the high scorers were disproportionately men—68 percent—probably because the test emphasized mathematics. Traditionally, American boys have gone further in math than have American girls. In addition, the top 5 percent were also lucky people—lucky because had they answered just one or two more questions wrong, they would have dropped down to being merely "bright." (This is a result of standard psychometric insistence on differentiating, described in chapter 2. When small differences are exaggerated, chance becomes all the more important.)

Who were in the bottom, the disreputable 5 percent? We pointed out that they were uneducated. In fact, 27 percent of the "very dull" had dropped out of high school at least *three years* before taking the test.[17] Furthermore, some of the bottom 5 percent had mental problems. Interviewers in later years rated forty-two of the white respondents as "mentally handicapped," and those respondents were likelier than others to be in the lowest 5 percent. If we believe that the AFQT is an IQ test, then all of the bottom 5 percent were mentally "retarded" by the conventional standard that IQ scores below 75 mark the retarded.[18] If we use a line of 70, then about one-fifth of the bottom 5 percent were retarded. (Unfortunately, the two ways of assessing mental handicaps—interviewer rating and AFQT score—hardly agree, casting more doubt on the AFQT as an intelligence measure.)[19] Including these two groups in the analysis distorts the picture of how intelligence operates in the general population, the huge majority of which is neither mentally retarded nor handicapped.

We also suspect that many of these low scorers were neither mentally disabled nor unintelligent but were instead discouraged test takers who gave up or were, to use the colloquial, "screw-ups"—respondents who rejected the test, fooled around, or just did not take it seriously. Why do we think that? Consider that sixty-nine of the whites (1.5 percent of them) scored *below chance*. Test takers who simply answered the questions randomly would have, on average, gotten twenty-six correct.[20] How do we interpret someone scoring below chance? Many of the questions were designed to be answered correctly by virtually everyone. Yet these test takers did worse than throwing darts at the answer boxes. We suspect that many of the respondents who scored below chance simply lost interest, perhaps because it was hard for them or for other reasons; others did not take the test seriously, either answering randomly or not answering many questions at all.

We believe that many of the lowest 5 percent, including many who scored below chance, were "screw-ups" for three major reasons. First, some of these low scorers had scored at or even above average in previous, school-given IQ tests.[21] Second, the below-chance scoring is greatest in the last section of the test: Three of the AFQT subtests appeared early in the test form (sections 2, 3, and 4). From 1.5 to nearly 5 percent of whites scored below chance on those. In the last subtest Herrnstein and Murray used, section 8 (which appeared after respondents had already faced *264* multiple-choice questions), *10* percent scored below chance—an indication to us that, as the test went on, more test takers "dropped out." On both sections 4 and 8, five respondents actually scored zero—a virtually impossible result on a lengthy multiple-choice test, suggesting that they did not even attempt the questions. The third reason we believe that many test takers dropped out is our own observations of test-taking among youth. This pattern of taking tests has wider implications.

The setting in which the NLSY respondents took the AFQT during 1980 seems to have been optimal for valid testing. About 700 of the subjects (6 percent) were tested individually. The rest were tested in groups of roughly ten in hotels, libraries, or similar locations. Two interviewers administered and proctored each group administration. Respondents were paid $50 (worth over $90 in 1995 currency) to show up; how they scored on the test did not matter. Respondents were also enticed to participate by being offered the test results and vocational information. The test takers were relatively mature as test takers go (fifteen to twenty-three years old) and had already demonstrated their commitment to the NLSY by being interviewed a couple of years earlier.[22]

Contrast this situation to the typical setting for standardized tests in schools: A teacher hands out booklets to a class of twenty to thirty-five restless, perhaps resentful, adolescents or preadolescents. The teacher then tries to maintain order during the long test. It is reasonable to assume that at least as high a proportion, but probably a higher one, of students become distracted or discouraged or just "blow off" that test as did the AFQT in the NLSY. Even one-on-one testing in schools is problematic. School psychologists in inner-city schools, at least, are pressed to do so many evaluations that their testing tends to be rushed and superficial.[23]

Students who ignore or resist teachers' instructions for taking a test are likely to do poorly in school, and they are likely to do poorly outside of school as well. Employers do not appreciate such rebellious attitudes any more than teachers do. Consequently, low scorers will often become low achievers. Perhaps this outcome is due, in part, to poor intelligence. But it is probably due to poor attitudes—one reason that such tests predict well even if they do not measure intelligence well.

Rebelliousness, apathy, and other attitudes reduce test scores. So do emotional conditions such as anxiety.[24] Where do such attitudes and feelings come from? Perhaps they are to some degree "natural," inborn temperaments. But many studies of youths point to social conditions—such as poor schooling, disorganized neighborhoods, stressed parents, high unemployment, and being in a stigmatized minority—that stimulate such attitudes. Once again, then, the evidence points to the environment as the basic explanation for test scores. We will revisit these points when we turn to explaining racial differences in test scores in chapter 8.

CONCLUSION

We have made several important critiques of using the AFQT as a measure of "natural intelligence." To these, Herrnstein and Murray can provide one rebuttal: The AFQT has predictive validity. That is, scores on the test are correlated with other, later outcomes that we care about. High scorers are likelier to go on to college, to do well in the military, to get good jobs, and so on; low scorers are likelier to fail in the same tasks. We grant all this (although how strong or weak the connection is with such outcomes we will see in the next chapter). But, as we pointed out in chapter 2, predictive validity tells us whether a test correlates with outcomes, but not *why* it does. Cognitive skills are surely part of the explanation for the correlation between tests and outcomes. Youths who read faster, who can juggle more numbers in their heads, who recognize more words, or who think more clearly will generally score higher on the AFQT. These skills will serve them well in school and, to a lesser degree, in life. One point is that such skills are learned in a variety of settings, some of which are more conducive to successful learning than others. The other point is that the AFQT in large degree reflects other characteristics besides some basic intelligence.

If those other characteristics were just "noise," just random influences on the AFQT scores, then they would not undermine *The Bell Curve* argument. But those other characteristics—such as how recently people were in school and self-defeating attitudes—are not mere noise. They influence how people end up in life. The problem is that Herrnstein and Murray attribute all the effect they have found of AFQT scores on outcomes to native intelligence; they should attribute much, if not most, of the effect of the AFQT to these other characteristics, such as instruction and motivation. Another way to understand what we have shown is that test takers' AFQT scores are good summaries of a host of prior experiences (mostly instruction) that enable someone do well in adult life. The error is to attribute the significance of the AFQT to some unseen native intelligence.[25]

The AFQT basically measures how much formal and informal instruction someone has received and absorbed. It is no surprise that youths who do well in school will usually continue to do well in school, in addition to doing well on the standardized tests. (Doing well on early tests also opens up more opportunities for advanced schooling, for example, by moving into a higher track.) And it is no surprise that doing well in school helps one to do well in life. As every parent of an adolescent is acutely aware, our society today is structured so that the college diploma is necessary even to imagine success. But this points to the direct consequences of *schooling* rather than of cognitive ability. (And cognition, unlike genes, can be changed by policy, as we shall see in chapter 7.) The AFQT also reflects mental disabilities, motivation, and attitude. And these traits, too, influence how well someone does in postschool endeavors—especially those endeavors requiring school-like skills and school-like discipline. For such reasons, the tests do predict outcomes well.

As *predictive* devices, then, the AFQT and similar tests work, at least in a general way, although they do not predict very well the performance of any given individual. But the AFQT predicts outcomes irrespective of whether it reflects intelligence or, instead, reflects other characteristics. For practical purposes—such as admission to college—predictive validity may be sufficient. But for *understanding*, predictive validity is not enough. We must be concerned with content validity, with what the tests actually measure, with why the tests correlate with outcomes. If the AFQT and similar tests essentially gauge an ability formed early in life, stable thereafter and fateful, then where people end up in life largely reflects natural differences and inequality is predetermined. If, however, these tests largely assess instruction and other environmental conditions, then inequality is neither natural nor fixed. The evidence strongly indicates that this key measure in *The Bell Curve*, the AFQT, is really a measure of instruction and attitude, of things that are changeable, not of innate intelligence.

The policy implications of this analysis are straightforward: Education—both formal and informal—matters. So do efforts to affect youths' motivations and attitudes. Instead of shrugging our shoulders and dismissing efforts to raise academic achievement, we can successfully invest in cognitive training (see chapter 7).

If the analysis of this chapter is correct, then *The Bell Curve* argument fails at this point. If the AFQT heavily reflects the contribution of instruction and social environment to individual achievement, then Herrnstein and Murray have not shown the dominance of native intelligence over environment, but the opposite. In the next chapter, however, we will assume that the AFQT does measure intelligence. We will evaluate how well the AFQT predicts outcomes compared with measures of the social environment. We will see that, even were we to scrap this chapter and grant all that Herrnstein and Murray claim about the validity of the AFQT, social context still better accounts for who gets ahead and who falls behind in the race for success.

Who Wins? Who Loses?

In THE OLD BLUES refrain, Albert King asks, "If you're so smart, how come you're not rich?" Herrnstein and Murray issue the academic equivalent of that challenge. Their evidence shows, they claim, that being intelligent leads people to be successful in school, wealthy, and stable in marriage; being unintelligent leads people to be poor, divorced, criminal, and even prone to injury. Herrnstein and Murray also claim that differences in people's social backgrounds are of minor importance in determining how their lives turn out. This is the crux of *The Bell Curve*'s empirical assertions, that intelligence more strongly shapes life outcomes than does social environment. The challenge to the book's critics, wrote Charles Murray in 1995, is the "way IQ dominates this thing we call 'socioeconomic status'" in explaining life outcomes.[1] *The contrary is true.*

In chapter 2 we showed that intelligence is not single, unitary, and fixed as *The Bell Curve* assumes. In chapter 3 we showed that, even if it were, intelligence is not well measured by the AFQT. In this chapter, using the very same evidence used in *The Bell Curve*, we show that even if the AFQT were a good measure of intelligence, the socioeconomic status of people's parents and their broader social environment more strongly determine life outcomes—specifically, who becomes poor.

We follow Herrnstein and Murray's data analysis as closely as possible and then go on, step by step, to correct their technical and conceptual errors. With statistics no more complex than theirs—just more accurate—we demonstrate that social contexts shape individual outcomes more than does intelligence, even if we grant them the dubious assumption that AFQT reflects intelligence. Later in the chapter, we step back to take a broader look at poverty and its causes. Redoing the survey analysis in *The Bell Curve* is a way to see the real dynamics that sort out winners and losers in America. Other scholars have also reexamined the NLSY, using somewhat different procedures than we did, and have come to the same conclusion: Herrnstein and Murray are wrong.[2]

Herrnstein and Murray exaggerated the importance of intelligence, not only by using a measure that largely reflects instruction, but also by defin-

Richard Arum coauthored this chapter.

ing social environment narrowly and incorrectly. For them, the educational attainment, income, and occupation of people's parents comprise the totality of their family environment. We, however, recognize and demonstrate that the schools, communities, regions, and social circles in which people live also shape their lives.

Herrnstein and Murray also exaggerated the relative importance of intelligence by making several technical errors in their analysis. We correct most of these. Ironically, one of their errors led them to underestimate statistically the importance of the AFQT score. But the rest of their errors led them to underestimate greatly the importance of social environment. (The bulk of our statistical analyses will be found in appendix 2. We have simplified the work for presentation in this chapter.)

By taking these steps to reanalyze *The Bell Curve*'s evidence, we not only refute Herrnstein and Murray, we accomplish a larger end: We show the many ways that people's ultimate fortunes depend on their social environments. This finding—that social context is more important than tested intelligence—has been repeatedly demonstrated in the social sciences over the years.[3]

WHO BECOMES POOR? AN OVERVIEW OF THE ANALYSIS

Which NLSY respondents became poor in 1990? Herrnstein and Murray set up a contest between two attributes to see which one most strongly predicts who was poor in 1990: the respondents' AFQT scores in 1980 or the respondents' parents' socioeconomic status (SES) around 1980. Herrnstein and Murray claim that the AFQT wins this match. The difference in the rate of poverty for people who scored high on the AFQT versus those who scored low on the AFQT is greater than the difference in the rate of poverty between those who scored high versus low on parental SES. On this evidence they conclude that "natural intelligence" is the key determinant of inequality. This result, the contrast between the strong association of AFQT with poverty and the weak association of parental SES with poverty, has become the central claim in Murray's defense of *The Bell Curve*.[4]

How could sociologists and economists have been so blind to the importance of intelligence? Perhaps they have not been blind but have been too scared to reveal this "truth." The answer is: They have been neither blind to it nor scared of it. Since the late 1950s an entire school of research has devoted itself to explaining economic outcomes. The 1972 book *Inequality* by Christopher Jencks and his associates is a well-known example of such studies. Social scientists have incorporated intelligence test scores—including specifically the AFQT—in their work.[5] These researchers have typically found that conventional measures of intelligence, including the AFQT, *are* correlated with important outcomes, especially with education and earnings. But the effects are not large; intelligence as measured in such tests does not have an overriding influence on economic outcomes; and test scores are but one factor among many that shape inequality.[6] This extensive and well-known research literature made us very skeptical of the claims that were emanating from press coverage of *The Bell Curve*. Years of accumulated research, not ideology, lies behind the academic community's chilly reception of *The Bell Curve*. On close examination of the book itself, our skepticism turned out to be well founded.

We redo the statistical analysis of the NLSY data in the following sequence:

- We review a few of the critical errors Herrnstein and Murray made and explain how we corrected them.
- We replicate Herrnstein and Murray's basic finding—that differences in AFQT scores matter more than differences in parental SES in predicting the chances that NLSY respondents were poor in 1990—and then show what the finding looks like once a few technical errors are corrected: The structure of respondents' families of origin is then about as important as AFQT score in predicting poverty.
- We expand the notion of social background to include aspects of the communities respondents lived in at the time of the testing and then show that these factors affected poverty, too. Together with home environment they are equally important as AFQT.
- We add into the explanation the formal educational experiences of the respondents, and at this point social environment broadly understood becomes significantly more important than AFQT as a predictor of poverty.
- We include two attributes of the respondents' communities in 1990 to show the importance of local conditions.
- We close the reanalysis by showing the critical importance to being poor of three factors that Herrnstein and Murray essentially ignored: being female, being unmarried, and being a parent. In the end, AFQT scores become one of the less important explanations of respondents' poverty.

In defining social background, Herrnstein and Murray overlooked some aspects of families, such as their size, and completely ignored other elements of the environment outside the family that nonetheless shape peoples' life chances, such as the kind of schools they attended and local job

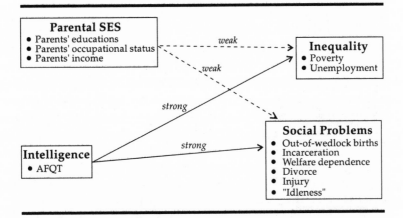

4.1. Herrnstein and Murray's Model of the Causes of Inequality and Social Problems (*Note*: Solid lines indicate strong effects; dashed lines indicate weak or insignificant effects. Herrnstein and Murray also control for age for technical reasons but do not examine its effects. *Source*: Authors' interpretation of *The Bell Curve*)

opportunities. Even our analysis does not capture nearly all of the social environment, because data are unavailable for many important dimensions of the environment—for example, on people's social contacts ("it's not what you know, but who you know").[7]

The authors of *The Bell Curve* seem blind to this broader meaning of the social.[8] They, and some of their critics, too, seemingly cannot raise their sights beyond the household to see obvious social influences on individuals. When, for example, a region enters a recession, many workers plunge into poverty—individual and family have little to do with it. For another example, residential segregation by race and class increase the risk of poverty for those who are isolated.[9] Yet Herrnstein and Murray's measures of social origin take none of such realities into account. As we will show below, the kinds of communities people live in substantially alter their chances regardless of their individual abilities or other personal traits.[10]

We can display the differences between the "natural inequality" explanation of individual outcomes and the environmental one we propose with the following graphs. Figure 4.1 shows what Herrnstein and Murray claim happens. People's genetically given intelligence, as measured in the NLSY by the AFQT, strongly determines their economic outcomes and thus determines their inequality. It also determines the chances that they become involved in problematic behavior. Herrnstein and Murray acknowledge

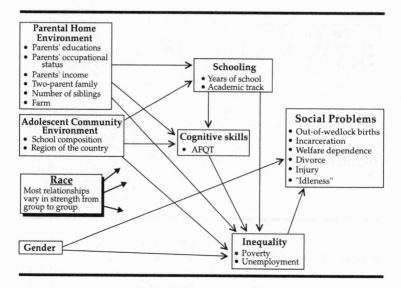

4.2. Our Model of the Causes of Inequality and Social Problems (*Note*: We have drawn in only the strong effects to simplify the diagram. We assume that parental home environment and adolescent community are correlated and adjust for that in our analysis, but we do not specifically analyze that correlation)

that people's family origins, specifically their parents' SES, has some effects, but those, they say, are weak.

Figure 4.2 displays our alternative. It is considerably more complex than figure 4.1 because the *reality* is considerably more complex. Original conditions include aspects of individuals' families (labeled here "Parental Home Environment") but also include aspects of the wider social milieux, such as the region people grew up in. Origins also include people's race and gender because these are genetic traits that take on *social* significance. (We do not in this chapter examine racial differences because we follow Herrnstein and Murray's procedure of examining inequality among whites only. Chapter 8 is devoted to the subject of race and ethnicity. We note for the record in figure 4.2 that racial identity influences most processes.) People's home and community origins affect the amount and kind of schooling they receive and, together with schooling, affect their cognitive skills—for which we use the AFQT as a measure.

Figure 4.2 is simplified for easier legibility by not displaying aspects of individuals' *contemporary* situations: their adult community environment, including where they live and the local unemployment rate; and their contemporary family situations, including whether they are married and have children. These conditions also affect the odds that people will become poor, independent of the other factors in the model. They would appear roughly in the center of figure 4.2, at the risk of considerable visual clutter. We do include them in the analyses that follow.

Adolescent environments, schooling, and contemporary environments in turn shape people's economic circumstances. Finally, those circumstances affect the odds that people will encounter social problems. We recognize that figure 4.2 is still a simplification. Aside from not displaying the current environment, it leaves out other social factors such as people's peer groups. Also, elements listed in each box are limited by what is available in the NLSY. For instance, we have no direct measure of school quality. Figure 4.2 also simplifies the complex causality. For example, getting injured or divorced often leads people to become poor. Still, the figure serves to contrast our perspective with the vastly oversimplified and mistaken one presented in *The Bell Curve*. It also provides a road map to the analysis that follows.

Fixing the Errors

Before we could use the NLSY to explore the sources of inequality, we needed to correct several errors in Herrnstein and Murray's analysis. One technical error actually led Herrnstein and Murray to *under*estimate how much AFQT score affected the risks of being poor. We corrected it in our analysis.[11] Yet the bulk of their errors led them to exaggerate the relative importance of AFQT score by greatly underestimating the importance of social environment. Most importantly, Herrnstein and Murray erred in how they constructed the measure they call "parental SES," the measure they use as the key—essentially the only—indicator of individuals' social environments. We quickly review four kinds of errors that they made.

MISSING INFORMATION

Key information was missing for many of the respondents. For example, 17 percent did not list their father's occupation. And 21 percent provided no information on their parents' income for either 1978 or 1979, about two-thirds of those because they were not living with their parents then, so they were not even asked their parents' income.[12] Herrnstein and Murray still used those respondents in their analysis, in effect assigning each of them the *average* parental income reported by the other respondents.[13] But these respondents with missing information were not average respondents.[14] By assigning them the average income, Herrnstein and Murray rendered their measure of parental income less accurate and thus less likely to show up as an important influence on becoming poor.[15] We corrected this error by using a more appropriate procedure for handling the cases with missing information.[16]

RELIABILITY

The term "reliability" refers, roughly, to how stable and trustworthy a measurement is. A metal ruler will give us the same measurement of an object time after time; it is reliable. A ruler made of soft rubber will give us varying measures of the same object; it is not reliable. In this research, the AFQT is more reliable than the measure of social environment. *Whatever* the AFQT measures—intelligence, years of school, motivation, and so on—it does so reliably. It gives roughly the same score for the same person time after time. That is because many years of psychometric work have gone into making it reliable (see chapter 2). The components of the parental SES scale tend to be less reliable. For example, youth are only approximately accurate in estimating their parents' income.[17] And only 4 questions are used to measure SES, compared with the 105 for the AFQT.

The implication of this difference in reliability between the AFQT and the components of parental SES is this: When two attributes are both equally associated with a third, the one of the first two that is more reliably measured will have a higher statistical correlation than the other. That is because its measurement contains less random error. So, in the face-off Herrnstein and Murray set up, the AFQT has the critical advantage of much greater reliability over parental SES. University of Minnesota public policy professor Sanders Korenman and Harvard sociologist Christopher Winship repeated Herrnstein and Murray's analysis, using a statistical adjustment to correct for unreliability in the AFQT and the parental SES index. By doing so they *reversed* a number of Herrnstein and Murray's conclusions: Corrected for unreliability, parental SES was often more important than the AFQT.[18]

This is one correction we did not do in our analysis, partly because others had done it, partly because not all scholars agree on its appropriateness, and partly because we have more substantive concerns. Had we done so, however, our own reversal of Herrnstein and Murray's conclusions would have been even stronger than the one we present below. In this decision, as well as others, we are more cautious in our critique than we might justifiably be.

WEIGHTING THE COMPONENTS OF PARENTAL SES

Herrnstein and Murray discussed in detail their construction of the parental SES index, proudly announcing that they bent over backward to build a good measure of social class background to run against the AFQT. But they did not. They added together mother's education, father's education, the head of household's occupation, and parental income, averaged for 1978 and 1979, to create an index. In principle, each of these four counted equally in building the scale. In practice, they made errors that let the two education measures count for more than occupation and income.[19] But their real error was to even try to construct such an index.

Here is why: In creating their four-component index of parental SES, Herrnstein and Murray were in effect insisting that each component determined outcomes with equal weight. For example, having a mother with a poor education, they tacitly assumed, should be as much of a risk factor for poverty as having parents with a low income. But what if this assumption is wrong? What if having a father or a mother with little education does not alter people's chances of becoming poor, but having parents with a low income substantially increases those chances? Lumping the different measures together in a scale would wash out the effect of parental income. This is precisely what happened. In the NLSY, how much money a respondent's parents earned makes a big difference in his or her chances of being poor, but how much education their parents had makes no difference. Herrnstein and Murray's parental SES index leads them to underestimate the importance of parental home environment, specifically income, for the risk of poverty.

We corrected this error by simply using each component separately in our statistics. There is no statistical necessity for combining measures into indices, and there are often good reasons for looking at the measures separately. That is how we learned that parental income is so important in explaining who becomes poor. (Even here our reanalysis leans in Herrnstein and Murray's favor. Income earned over two years, even if accurately reported, is an unreliable measure of a family's affluence. Economists prefer to look at measures such as accumulated wealth or income received over several years.)

OMITTED VARIABLE BIAS

The fearsome label "omitted variable bias" refers to leaving important causal factors out of an explanation. Such omissions mean that one both misses the whole story and distorts the analysis of the causes that are in the explanation. The bulk of our corrections to *The Bell Curve* analysis concern this problem: Herrnstein and Murray left out many important features of the social environment that affect who is at risk of being poor. We added them back in.

For example, in trying to estimate the social class background of each respondent, Herrnstein and Murray included income (sort of; see above), but they did not include the number of siblings each respondent had. Many studies show that the more siblings people have, the lower their families' effective wealth and the lower their own chances of getting ahead. People who grow up with no or one sibling get more space, resources, and attention than those with several siblings.[20] We included the number of siblings each respondent had had in 1979 in our reanalysis. We also included whether or not the respondent had been reared on a farm, because calculations of income and needs differ for farm families (see next box), and whether the respondent had grown up in a two-parent family, because that strongly affects a family's long-term wealth.

The Bell Curve also ignored the community context within which people were raised and within which they currently live. Whatever the home situation from which people come, the communities in which they grow up and the communities in which they currently live also shape their fates, the first, for example, by providing schools of varying quality, and the second, for example, by providing jobs of varying quality. Herrnstein and Murray ignored these sorts of environmental conditions; we included them, and they made a difference.

Researchers do make errors; in any project as large as *The Bell Curve*, they are bound to happen. Unfortunately, the great bulk of the errors Herrnstein and Murray made led them to underestimate how much the social environment changes the odds that someone will become poor. And that underestimate, in turn, made a low AFQT score seem more important in determining who becomes poor. We corrected these (and other) errors in our reanalysis; consequently we came to more accurate conclusions. Compounding the problems of measuring parental home environment and adolescent community environment are the problems of how to interpret the AFQT scores. As we argued in chapter 3, the AFQT captures a broad array of instructional experiences that are related closely to both periods of youths' environments. In other words, some of the social influences on outcomes are indirect via their effects on the AFQT. To the extent to which that is true, Herrnstein and Murray also overstate the role of intelligence in the AFQT effects.

DEFINING POVERTY

Deciding who should be counted as poor is a complex task. The federal government ties the definition of poverty to "nutritional needs" as they were understood in the early 1960s. In 1968 a group headed by Dr. Mollie Orshansky of the U.S. Social Security Administration took the Department of Agriculture's calculations of how much it cost to meet the nutritional needs of a family of two adults and two children. They adjusted that number for other combinations of adults and children and also took into account that farm families could contribute their own produce. Because surveys at the time showed that poor people used one-third of their incomes for food, Orshansky's group multiplied the cost of food by 3 to estimate families' total needs. The result was a table that defined the "poverty lines" for farm and nonfarm families of varying sizes. Since 1968 the Census Bureau has adjusted each of these poverty lines for changes in overall prices, but not for changes in the prices of the specific foods. Nor have changes in nutrition, tastes, or other eating patterns been taken into account. Nor has there been any attempt to check the assumption that the ratio of total needs to nutritional needs is 3. Most researchers realize that the poverty line understates national rates of poverty; some prefer to classify families with incomes 25 percent over the official line as "poor."* The original poverty line for a nonfarm family with two adults and two children was an annual income of $3,477 (in 1968 currency); adjusting for inflation yields an official poverty line for a similar family in 1994 of $15,029.

* See, e.g., Jencks, "Is the American Underclass Growing?"

Poverty, Test Scores, and Parental Home Environment

Herrnstein and Murray's key statistical analysis compares how AFQT scores and parental SES influenced the odds that an NLSY respondent was poor in 1990. We have already shown (in chapter 2) how Herrnstein and Murray got from the number of correct answers on the AFQT to their zAFQT score of intelligence. And we have just discussed the problems with their measures of parental SES. Our measures of parental home environment include siblings, farm residence, and whether the respondent lived in a two-parent family at age fourteen. These together assess family background more accurately and more powerfully than does Herrnstein and Murray's flawed index. This set also shows how much more complex people's family situations are than Herrnstein and Murray acknowledge.

(Korenman and Winship, in their reanalysis, went farther than we do, including home attributes such as whether the respondent's mother worked, what age she was when the respondent was born, and whether the family regularly received newspapers. The reader can with little effort think of several other features of home environments that shape children's lives but were not even asked about in the NLSY—for example, parents' work schedules, family wealth, parents' physical and mental health, and the involvement of other relatives.)[21]

We begin our reanalysis by reproducing Herrnstein and Murray's results concerning the chances that an NLSY respondent was poor in 1990 (see details in appendix 2). Then we contrast that pattern with what we get using the corrected measures of parental home environment. We follow Herrnstein and Murray's practice of graphing the results of the statistical analysis as a method of making comparisons.

Part (a) of figure 4.3 reproduces *The Bell Curve*'s key finding (p. 134) as closely as possible.[22] Here is how to read the graph: The vertical axis shows the probability that an NLSY respondent was poor in 1990. The horizontal axis displays variation in any particular causal variable, from −2 "standard deviations" below the mean (which is about where the bottom 5 percent of people are) through the mean to +2 standard deviations above the mean (where one finds the top 5 percent or so).

The solid, higher line in figure 4.3(a) shows the association between respondents' AFQT scores in 1980 and the probability of being poor in 1990, for respondents with average parental SES as Herrnstein and Murray measured it and average ages (within the range of 25 to 34). Respondents who scored −2 standard deviations below the mean on the AFQT had an estimated probability of being poor of about 28 percent. That is, respondents of average parental SES and average age who scored that low on the

On opposite page:

4.3. Probability That an NLSY Respondent Was Poor in 1990 by AFQT Score and Family Background: (a) Herrnstein and Murray's Calculations; and (b) After corrections (*Note*: All variables except either parental home environment—dashed line—or AFQT score—solid line—were set at their mean values. Correction involves adding number of siblings, farm residence, and two-parent household to the parental SES index to form the parental home environment index, and also replacing uniform weights for index items with "effect proportional" weights. *Source*: Authors' analysis of NLSY data)

The comparative effects of AFQT scores and parental home environment on the probability that young white adults are poor: Herrnstein and Murray's original results show AFQT scores to be much more important, but corrections show that they are not.

(a) Herrnstein and Murray's Calculations

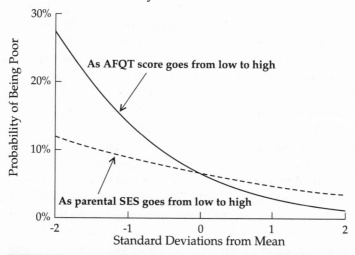

(b) Corrected Measures of the Effects

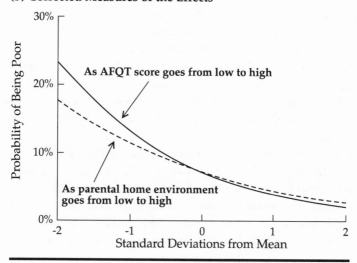

AFQT had a .28 chance, better than one in four, of being poor. That probability drops sharply to between 7 and 8 percent, about one in thirteen, for respondents with average AFQT scores and then drops to about 2 percent, a one-in-fifty chance, for very high scorers +2 standard deviations above the AFQT mean. The lower, dashed line shows the same information for parental SES—how the risk of being poor varied for persons with average AFQT score and age. Respondents who had parents of very low SES had a 14 percent chance of being poor in 1990, while those who had very high SES parents had about a 5 percent chance of being poor. Note the contrast in the lines: The solid AFQT line is much steeper than the dashed parental SES line. There is a twenty-one-point drop in the risk of poverty from very low to very high AFQT score versus only a nine-point drop from very low to very high parental SES. This is the basis of the trumpeted claim that intelligence "dominates" SES as an explanation of poverty.

(These statistics are derived from "logistic regression analysis." For a brief overview of regression analysis, see appendix 2.)

But now let's see what happens after making a few elementary corrections. Part (b) of figure 4.3 shows the consequence of correcting the measure of parental SES and also expanding it to be a fuller measure of parental home environment. Separating the SES components into the two parents' education, occupation, and income;[23] correcting the treatment of missing data; and adding siblings, farm residence, and whether there were two parents in the respondents' homes—substantially increase our estimate of how much family origin influenced the chances of becoming poor. It influenced the risks of poverty about as much as AFQT scores did. In quantitative terms, the corrected estimate of the family effect is 86 percent as large as the corrected AFQT effect (the AFQT effect also increased a little because we corrected a technical error Herrnstein and Murray made). Comparing extremes, we see that, all else constant, 18 percent of people from the poorest families grew up to be poor while only 4 percent of people from the most advantaged families became poor. For technical reasons, even this correction underestimates the importance of home environment compared with AFQT score.[24]

Murray challenged critics of *The Bell Curve* to show how, in any fashion, parental SES can be as important as AFQT score.[25] We have shown that simple corrections of errors suffice to meet that challenge. Even a limited measure of home environment is as important a predictor of poverty as the AFQT measure is. Korenman and Winship also conclude, from their reanalysis of the data: "Estimates based on a variety of methods . . . suggest that parental family background is at least as important, and may be much more important than [AFQT score] in determining social and economic success in adulthood" (see also work by Dickens et al.).[26]

Poverty and Communities

People's social environments involve more than simply the financial, educational, and demographic assets of their own families.[27] The local community matters, too. Research shows, for example, that women's chances of getting married depend on the number of men in the area who are employed at good wages.[28] The immediate neighborhood also affects people's ways of life, whatever the family's own resources. This simple fact is one reason why Americans try so hard to find and afford "good neighborhoods."[29] It is one thing to come from a low-income family but live in a pleasant suburb with parks, low crime, and quality schools, and another thing altogether to live in an inner-city neighborhood that lacks those supports. The child in a well-endowed community gets the benefits of the locale regardless of his or her family's particular situation, just as the inner-city child bears the burdens of a low-income community even if his or her family might have a moderate income.

Social scientists have increasingly taken community context seriously and have found compelling evidence that residential segregation and the concentration of the disadvantaged exacerbates the consequences of poverty, family break up, crime, and deterioration.[30] (In chapter 8, we will see the role of residential segregation on minority achievement.) The concentration of the disadvantaged in particular communities and particular schools undermines the fortunes of otherwise able youth. Schools in low-income and minority neighborhoods tend to lack resources and quality instruction; the concentration of children with problems can distract from learning; and separation by social class cuts poor children off from friendships with advantaged ones. In the local neighborhoods, similar effects occur. Low-income areas have fewer jobs, fewer resources, and poorer-quality services than do affluent ones. Local culture can clash with high aspirations, public disorder creates insecurity, and remaining opportunities leave.[31] Other youth with similar personal liabilities and similar disadvantaged families but who find themselves in good schools and neighborhoods benefit from the obverse of all these conditions.[32]

From the NLSY, unfortunately, we could extract only two measures of the respondents' adolescent community environments. One, their region of

83

residence in the first year of the survey (1979), is quite global but captures some differences in economic and cultural conditions. The other is quite specific. It is an index of *school composition* for the high school the respondent last attended. The index is composed of three indicators: the percent of the student body that was nonwhite, the percent that was economically disadvantaged (that qualified for free lunches), and the dropout rate.[33] (Direct measures of instructional resources and quality were unavailable.) This measure tells us about the immediate context in which the respondents spent their days as teenagers and, because schooling is locally organized, tells us something about the local community as well. The higher the score—the fewer dropouts, the fewer poor students, and the fewer black students—the more advantaged the communities the respondents grew up in. (We do not mean to suggest that low-income and black students are per se drags on achievement. The fact is that in the United States schools and neighborhoods where poor and minority youths are concentrated tend to lack all sorts of "social capital."[34] Also, poor students tend to have problems they bring from home, and that contributes to a social climate that may interfere with attainment.)

These two partial measures of adolescent environments make a statistically important difference in the respondents' probabilities of being poor. Since we have "controlled" or "held constant" attributes of the *individuals*—notably their AFQT scores—these results show that the wider *community*, as well as the specific family, shape individuals' economic fortunes. The combined, distinct effects of family of origin plus community of origin equal the distinct effect of the 105-question AFQT score. As figure 4.4 shows, the poverty rates predicted by parental home and adolescent community environments, on the one hand, and by AFQT scores, on the other, are similar.[35]

The effect of communities is even greater than appears in figure 4.4, because local conditions can temper or aggravate the consequences of getting low scores or coming from poor families. For respondents with average AFQT scores and average parental home environments, the poverty rate if they had attended disadvantaged schools was 14 percent, over ten points higher than for identical respondents who had come from advantaged schools, 4 percent.[36] But the contrast by school composition is even greater for those who had low scores or poor parents, about fifteen to twenty points difference in the chances of becoming poor depending on whether the school was above or below average in composition.[37] Disadvantaged communities put even relatively advantaged individuals at risk; advantaged communities help even the unfortunate. Disadvan-

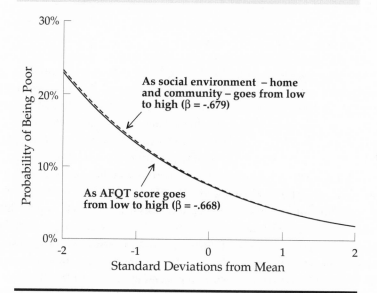

Adding adolescent community environment – school composition and region – to home environment shows that social environment and AFQT scores have equal effects on the probability that young white adults are poor.

4.4. Probability That an NLSY Respondent Was Poor in 1990 by AFQT Score and Social Background (*Note*: Social background includes parental home environment and adolescent community environment. For computing the plot, all variables except either social background—dashed line— or AFQT score—solid line—were set at their mean values. *Source*: Authors' analysis of NLSY data)

taged communities have schools where learning is hindered, they lack the economic activity necessary to provide jobs, and they even limit the chances of marrying out of poverty. Advantaged schools help residents in these and other ways.[38] These calculations make the fundamentally sociological point that individuals' fates are not theirs alone. Their life chances depend on their social surroundings at least as much as on their own intelligence.

We could stop right here because, by repeating *The Bell Curve*'s analysis, we have shown that *social environment during childhood matters more as a risk factor for poverty than Herrnstein and Murray report and that it matters *statistically at least as much as do the test scores* that purportedly measure intelligence. (Recall that we have been conservative in our re-analysis.) The key finding of *The Bell Curve* turns out to be an artifact of its method. Although we could rest our case against *The Bell Curve*, we will go on, because our larger purpose is to explore the social sources of inequality.

Poverty, Test Scores, and Education

In chapter 3 we showed that the AFQT largely measured schooling (and motivation), but for the purposes of this chapter we have adopted Herrn-stein and Murray's assumption that the AFQT measures intelligence. If so, what is the role of schooling itself in determining the chances of being poor? Social scientists have long established that people's educational at-tainments are the strongest immediate determinants of their economic for-tunes.[39] The further you go in school, the more money you make. Educa-tion works in part by "transmitting" the effects of earlier experiences. That is to say, how much education people obtain itself depends in part on the affluence and education of their parents. But how much education students attain is not the result *only* of such personal factors. Schools themselves and what happens in them—the instruction, the teachers, fellow students, and so on—also determine how much students learn and how far they go (see chapter 7). Independent of talents or personal background, then, schooling is part of the social environment.

We now add formal schooling into our analysis. In estimating the effect of education on poverty we distinguish between two phases of education—the years of school respondents had completed *before* taking the AFQT and the years completed *after* taking the AFQT. We also distinguish between an academic high school education and a general or vocational one.

Figure 4.5 shows how including education further accounts for the chances that someone became poor. At this stage in our reanalysis, formal schooling and childhood environment are each more important than AFQT score in predicting the risks of poverty. If one takes the AFQT as a measure of intelligence, then figure 4.5 shows that intelligence does affect the odds of being poor. But it also shows that formal schooling and childhood envi-ronment matter more.

As we discussed in chapter 3, Herrnstein and Murray contend that how much formal education people obtain largely reflects their intelligence. One should, therefore, assign the effects of formal schooling in figure 4.5

86

Adding education to the analysis gives a more complete description of what determines the probability that young white adults are poor: education and environment more than AFQT.

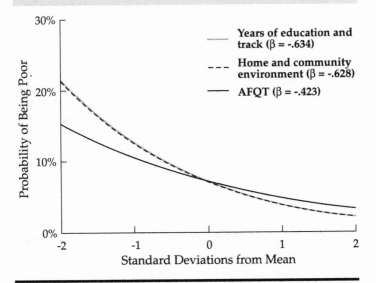

4.5. Probability That an NLSY Respondent Was Poor in 1990 by AFQT Score, Social Environment, and Formal Education (*Note*: Social environment here includes only parental home and adolescent community environments, not the contemporary environment; education includes years of schooling completed before taking the AFQT, years completed after, and whether the respondent had been in an academic track. For computing the plot, all variables in the equation were set at their mean values, except for, each in separate estimates: social environment (dashed line), AFQT score (black line), and education (gray line). *Source*: Authors' analysis of NLSY data)

to intelligence. But we showed that, for a few reasons, it is much more likely that the opposite premise is true, that test scores are the *consequence*, not the cause, of formal education. We should then interpret the effect of the AFQT in figure 4.5 as largely representing the effects of those aspects of schooling not captured by years or track—for example, the quality of teaching, the content of courses, and extracurricular instruction. More conservatively, we could claim instead that the AFQT line in figure 4.5 shows the effects of variation in respondents' verbal and math skills not attributable to their formal schooling. Whatever the interpretation, such skills are less important than formal schooling and childhood environments in affecting the risk of becoming poor.

Adult Community Environment

To this point, we have largely looked at how young people's social environment in adolescence influences their chances of becoming poor in adulthood. But the contemporary community context surely matters, too. Most important, perhaps, is the local economic situation. The chances of being poor are higher for residents of communities in the economic doldrums. The only direct measure we have of that in the NLSY is an approximate indicator of the unemployment rate in the respondents' labor markets.[40] We also can distinguish whether respondents lived in inner cities, suburbs, or rural areas. Together, these two indicators, rough as they are, further explain who became poor in 1990 (data not shown; see appendix 2).[41] Finer geographical distinctions, by neighborhood perhaps, and further descriptions of contemporary communities would no doubt have shown yet stronger effects. But the key point for us is that local economic conditions—far beyond the control of any individual—help determine who becomes poor and who does not. If jobs depart, more people, smart and dumb alike, become poor.

Poverty, Gender, and Adult Family Environment

Perhaps the most surprising omission of all in *The Bell Curve*'s discussion of poverty is any recognition that women are far likelier to be poor than are men. Figure 4.6 shows how great the gap was in the NLSY. It compares men and women by their AFQT scores, holding constant age, education, and social environments. A young woman would have had to score *forty-one points higher* on the AFQT than a young man of the same age, formal schooling, and background in order for her risk of being poor to have been as low as his. Put another way, just being a woman raised a respondent's risk of poverty by the equivalent of forty-one AFQT points (the difference between being "dull" and being "bright"). A similar point can be made concerning parental income. Holding the AFQT and the other factors constant, a woman's parents would have had to have earned $63,000 more than

A white woman needs to score 41 AFQT points higher than a white man from the same environment and with the same education in order to reduce her probability of being poor to his lower level.

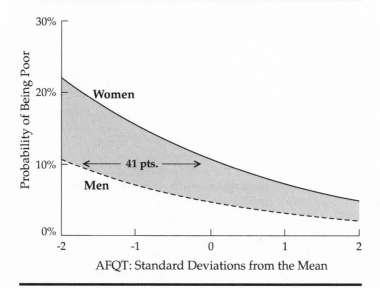

4.6. Probability That an NLSY Respondent Was Poor in 1990 by AFQT Score and Gender, Other Variables Held Constant (*Note*: For computing the plot, all variables except AFQT score were set at their mean values. *Source*: Authors' analysis of NLSY data)

the parents of an otherwise similar young man for her to have a risk of poverty as low as his.

Why should a woman be at so much greater risk than a man with similar qualifications and family advantages? To understand the answer to that question is to understand some fundamentals about poverty in the United States. We will turn to those fundamentals later. Suffice it to say for now that women's low wages and high family responsibilities play important independent roles. Before that, we turn to the issue of marriage.

Marriage affects the risk of poverty more than does any other attribute we examined in the NLSY. Table 4.1 shows the estimated rates of poverty in the NLSY for white men and white women who were identical to one

TABLE 4.1
Poverty Rates Are Much Higher for Women,
the Unmarried, and Parents, When Other Attributes
Are Held Constant

Family Status	White Women	White Men
Total	11%	3%
Unmarried	45	23
Childless	30	13
Two Children	64	40
Married	4	1
Childless	2	<1
Two Children	8	2

Note: These rates are predicted by model A8 (appendix 2) for white high school graduates with average AFQT scores, parental home, and adolescent community environments, residing in central cities of metropolitan areas with average unemployment in midwestern states.

another in AFQT score, parental home, adolescent community, and current adult environments, and schooling. Being married reduced the chances that a man was poor from 23 in 100 to 1 in 100. Being married reduced a woman's odds from 45 in 100 to 4 in 100. The basic reason, of course, is that married people can draw on two wage-earners. A closer look shows that, for respondents from advantaged parental and adolescent backgrounds, marital status made a minimal difference; few of those people became poor, married or not. For respondents with below-average backgrounds, however, being unmarried raised the risk of poverty greatly. Just being a woman increases the risk of poverty; just being unmarried does, too; being both is a double-whammy. (Herrnstein and Murray argue that being unmarried is itself a result of low intelligence. Thus, they would probably claim that marriage's effect on the risk of poverty should be attributed to differences in intelligence. But married respondents scored no higher on the AFQT than did unmarried ones.[42] Therefore, the heightened risk of poverty that the unmarried experience has little to do with their intelligence; it has to do with the need for two incomes.)

Having children also dramatically increases the risks of being poor. Two-thirds of the white women in the NLSY *with high school diplomas*, supporting two children and not currently married (even though they may have been married when they had the children), were poor. The presence of children also magnified the effect of social origin. Unmarried mothers from affluent backgrounds were not likely to be poor, but coming from a poor background meant that the consequence of being unmarried, or of having children, or the combination of the two was disastrous.

Why are women so much more likely to be poor than men? Low wages and, for those who are mothers, few hours of work account for most of women's poverty. Although many women gained significant ground on men in the 1980s, most of those gains came to full-time workers, older workers, and college graduates.[43] Many of the young women who first became mothers in the 1980s were not part of that change. Married women are not as poor as unmarried women simply because they share in their husbands' higher incomes. In the end, men are not as likely to be poor as women are because they earn higher wages, work more hours, and have fewer child-care responsibilities.

This pattern needs to be seen in historical perspective. The high poverty of unmarried women, especially of unmarried mothers, comes from the combination of economic need and economic disadvantage that is peculiar to the United States in the past twenty-five years. Other societies, even this one in earlier times, have shielded young mothers and their children from the risks of the marketplace. Contemporary American society is unique in its lack of provision for children (see chapter 6). Here, then, is a major factor that determines who among adults is poor: gender. Its effect cannot be the result of natural intelligence, since men and women score equally on "IQ" tests (indeed, the tests were constructed to yield no gender differences). It cannot be the result of a "free" market since such a market presumably cares little whether a worker is male or female. It is a result of how American society, from schools through jobs, is structured and of how our policies operate.

From this feminized poverty, child poverty follows. Because most children live with their mothers (or, if not, with female kin), they share their mothers' vulnerability to few hours at low wages. Calls, such as those in *The Bell Curve*, to solve women's and children's poverty by getting single mothers married are too simplistic. They assume that each poor mother could marry a man able to earn at least enough to keep a family out of poverty. The actual incomes single American men make are not sufficient to keep all single mothers out of poverty even if we could pair them off. And men's earnings have been declining. Single men are disproportionately poor, too. By our calculation, 23 percent of average single men in the NLSY sample were poor (table 4.1); those from disadvantaged family or school backgrounds were even more likely to be poor.

A Note on Race

We have kept the focus here on explaining economic inequality among whites only, both to replicate *The Bell Curve* and to simplify matters. However, we repeated these analyses for the African Americans in the NLSY, and the results are essentially the same (see appendix 2). We have also found, as others have, that scores on the AFQT seem to account (in the statistical sense) for the black-white difference in the risk of being poor among the NLSY respondents. Explaining that racial difference in AFQT scores is the burden of chapter 8.

Summary

Herrnstein and Murray built their case that intelligence determines inequality by trying to show that whether or not an NLSY respondent was poor was more strongly correlated with the respondents' AFQT scores than with their parents' socioeconomic status. In this chapter, we accepted Herrnstein and Murray's evidence, their measure of intelligence, and their basic methodology and then reexamined their results. By simply correcting a handful of errors, we showed that coming from a disadvantaged home was almost as important a risk factor for poverty as a low AFQT score. By including in the analysis simple attributes of respondents' communities of origin, we showed that a respondent's social background—family plus local community—was just as important as AFQT score in predicting the likelihood that someone became poor.

Moreover, respondents' formal *schooling*, which Herrnstein and Murray effectively excluded from their analysis, turned out to be—as social scientists have long known—critical in explaining who became poor. In comparing formal schooling, adolescent environments, and the AFQT, the last one comes in third in the "contest" to explain who was poor in 1990.

Even more critical to identifying who became poor—and totally ignored in *The Bell Curve*—is gender. All else equal, female respondents were far likelier to be poor than were male respondents. And yet more critical still were marriage and parenthood: The unmarried were likelier to be poor than the married because they subsisted on one income. And respondents with children were likelier to be poor than the childless because children increase financial needs and restrict women's ability to work. By the time all the pieces of the puzzle are put into place, low intelligence—assuming that all the variation in AFQT score is due to intelligence—is revealed to be relatively far down on the list of risk factors for poverty.

Even more meaningful than debunking *The Bell Curve*, this statistical work shows how inequality *does* operate. Children *may* start off with different "natural" advantages useful for economic advancement (and such advantages probably include far more than just the sort of narrow intelligence psychometricians dwell on, advantages such as energy and good looks). But children *certainly* do start off with different social advantages, some with more parental resources and better conditions in their communities than others, and some with the advantage of being male. Children in better-off families and better-off places then receive better schooling and develop their cognitive skills further. Having good schooling and skills, added to the original advantages of gender, family, and neighborhood, combine with contemporary circumstances, such as being married and living in an economically booming area, to reduce substantially people's risk of poverty. Young adults who have lost out—in family advantages, in earlier community conditions, in gender, in schooling, or in current community conditions—suffer a heightened risk of poverty. This is a more complex explanation of who wins and who loses in the American race for success than is *The Bell Curve*'s "natural inequality" explanation, but it is the truer one.

In the next main section we leave the narrow, reductionist framework that *The Bell Curve* provides for understanding inequality and look more broadly at the issue of who becomes rich and who becomes poor in America.

Addendum: Explaining Incarceration and Unwed Motherhood

The Bell Curve claims that native intelligence also explains a variety of other outcomes besides poverty. We reexamined two of those outcomes: the odds that an NLSY man spent time in jail and the chances that an NLSY woman became an unwed mother. We found the intelligence explanation for them also wanting. (See details in appendix 2.)

Herrnstein and Murray argue that it was the unintelligent men in the NLSY study who ended up in jail. We corrected additional errors in their statistical work and replicated their basic finding. If all one considers is AFQT score and parental SES scale, then the probability of being incarcerated between 1981 and 1991 depended only on the former.[44] A fuller analysis reveals, however, that at least three other factors were jointly more important in predicting who ended up in jail: the schooling men had completed at the time they took the AFQT, the kind of high schools they last attended, and having been poor just before the 1980s. (Korenman and

Winship also find that social background factors, even excluding poverty, explain the chances of ending up in jail.)[45]

The Bell Curve argues, as well, that low intelligence explains which women had children out of wedlock. Again, we replicated their result.[46] Comparing the AFQT and the parental SES measure, only the former predicts who had an out-of-wedlock birth during the 1980s. However, a fuller analysis shows that social context is notably more important than AFQT score in accurately predicting who became an unwed mother. How much schooling a woman had completed at the time she took the AFQT, the number of siblings she had, whether she grew up with both of her parents, and, most important, whether she had been poor in any of the three years before 1981 largely account for whether or not she became a single mother. In the face of these conditions, the AFQT score is a rather minor risk factor.[47] (Again, Korenman and Winship draw a similar conclusion. They also examine additional *Bell Curve* outcomes and decide that the importance of intelligence was generally exaggerated.)[48]

In both the case of jailing and the case of unwed motherhood, low AFQT score is but one of several contributors. Social background and social conditions matter more. Most important is being poor, which, as we have seen through the bulk of this chapter, is largely the product of circumstances, not native intelligence. We return, then, to the subject of wealth and poverty.

RETHINKING WEALTH AND POVERTY

By looking at the big picture, we can now rethink the whole question of how people become poor. Most Americans get their money the old-fashioned way: they earn it by working at a job that pays them wages or a salary. Few families rely for their income on dividends, rent income, pensions, child support payments, or welfare. The implicit "rules" for getting money differ depending on whether it comes from wealth, earnings, or government checks.

Wealth is often inherited by one generation from another; that is one reason children of affluent parents are less likely than others to become poor. Yet only one-third of Americans with over $1 million income in 1989 began with an inherited fortune.[49] The typical first-generation millionaire in the late 1980s was an entrepreneur who got rich because the company he or she owned made a big profit. Starting a new business is the highest-stakes game in the U.S. economy; only a handful generate fortunes. Ray

Kroc's first hamburger stands were hard to distinguish from dozens of others in Southern California, but his McDonald's empire ultimately yielded so much corporate and private wealth that Kroc could afford to buy the San Diego Padres baseball team. Microsoft's Bill Gates accumulated one of the largest fortunes in America in less than ten years; he was able to buy Leonardo DaVinci's *Codex* for $30.8 million. But these are exceptions among a more modest group. Most entrepreneurs go bankrupt.[50] Those who don't usually earn only a middle-class living for their families.

Most American families, however, rely on the earnings of their adults—how many hours they work and how much they are paid per hour. How much they are paid, in turn, depends greatly on their schooling, gender, and race. How many hours they work depends a great deal on how many hours their employers ask of them. When employers ask for overtime, earnings increase; when only part-time jobs are available, earnings decrease. Many wage-earners have supplemental income from sources such as interest, alimony, aid, or cash they make in a garage sale.[51] Family income is the sum of the incomes of a family's members who live together.

Poverty is defined as having low income relative to the size of the family (see box, p.79). Because income varies far more from family to family in the United States than does family size, the main factor accounting for which families are poor is low income. Yet over time what makes a family's income drop below the poverty line is more often the addition or exit of a family member than a drop in income. For seven out of ten families that fell into poverty during the 1980s, a change in family composition precipitated the fall.[52] More often, that change was a divorce or marital separation; less often it was the birth of a child. A moment's reflection can make the importance of family change clear. A divorce or separation removes all or most of the man's income from the woman's and children's budget. The average man between ages twenty-five and sixty-five had an income of about $27,000 in 1993.[53] Very few unbroken families experience a drop of income that large in one year. Even when adults become unemployed, they collect unemployment compensation, find another job, or make up the lost wages in other ways. But when a man with an average income leaves a family, the members who remain lose all or part of his $27,000 income. Furthermore, they must rely on the earning power of the women and children who remain. In the United States, that earning power is substantially less than that of the man who moved out. (Alimony and child support provide only modest help.)[54] Meanwhile, the separated man sets up a new, one-person family with an income of $27,000 per person or moves in with another family, increasing their income by $27,000.

95

The overall United States poverty rate has hovered between 11 and 15 percent since 1969. That is because, in the aggregate, neither family incomes nor family size have changed much. But that figure masks two important facts about American poverty. The first is that the age composition of the poor has changed dramatically. The second is that there is great turnover in who is poor.

In 1966, 28 percent of Americans over sixty-five years old were poor; that dropped to 12 percent by 1994. During the same period, poverty among children rose from 18 percent to 22 percent (see figure 5.5). Improvements in older Americans' standard of living reflect the safety net of social security, coupled with the higher earnings that the recently elderly made during their lifetimes (compared with the people who were over sixty-five in 1969). Meanwhile, the poverty of children today reflects the decline in benefits to poor children together with increases in the number of single mothers. Children are more likely than ever before to live with their mothers but not their fathers. Those mothers earn lower wages in the job market than men do. In 1994 full-time employed women of prime working age earned 73 percent as much as men of prime working age. Also, social security benefits are adjusted for changes in prices, but the benefits for children are not. Between 1970 and 1993, the average social security check for a retired husband and his wife increased in purchasing power by 57 percent (from $729 to $1,145 in 1993 dollars). In that same period, the average Aid to Families with Dependent Children check for a family *dropped* in purchasing power by 46 percent (from $698 to $377 in 1993 dollars).[55] Thus, changes in government policy, as well as in American family life, partly account for why poverty among the elderly and poverty among children have moved in opposite directions. (See also chapter 6.)

The second important aspect of poverty hidden in the stable poverty rate is the turnover among the poor. Americans sometimes speak as if the "the poor" formed a group with fixed membership, different from the rest of us. That is inaccurate. Data from the Panel Study of Income Dynamics (PSID), a national survey that has tracked 5,000 American families since 1969, show that half of the people who are poor in any given year are not poor the next year. Only 8 percent of the poor—fewer than 2 percent of all Americans—have been poor for as long as three years. Incomes and family circumstances change so much that there are more border crossings between poverty and nonpoverty than between the United States and Canada.[56] That is both good news and bad news. The good news is that very few families are poor for long. The bad news is that the 11 to 15 percent of American families that are poor at any given moment are the tip of a poverty iceberg. In 1991, 14 percent of Americans were poor in an average month, but 20 percent were poor for at least two months during the year.[57] Of the over 5,000 families the PSID has tracked, 31 percent have been poor for at least one of the twenty-five years of the study. It would not be too much of an exaggeration to say that "We have met the poor and they are us."

These brief notes about poverty point us toward a deeper sociological and economic understanding of financial success and failure. We gain only a modest comprehension by focusing on people's intelligence, because that does not help explain the rising fortunes of the elderly, nor the declining fortunes of children, nor why unmarried women are at such a higher risk of poverty, nor why neighborhoods and economic fluctuations affect the risks of poverty so much. We are pointed instead toward considering economic cycles, family changes, community differences, and government policies. These are more powerful and more enduring determinants of poverty in our society. For individuals, as we argued with our sociological model, success or failure depends heavily on where they are located socially, geographically, and historically when they grow up, when they enter the job market, and changes in life circumstances afterward, such as parenthood and divorce. The fate, for example, of an Appalachian widow in the 1930s was vastly different from the fate of a young man with a master's of business administration in the 1980s, a much greater difference than any imaginable gap in their genetic endowments or intelligence could ever account for.

CONCLUSION

If what we want is talent to be rewarded, intelligence may be *too* small a factor in determining economic inequality in the United States. We have reanalyzed *The Bell Curve*'s evidence on who finishes last in the American economic race. Lack of native intelligence, as Herrnstein and Murray define it, may contribute to a racer's chance of finishing poorly. But as or more important is each of several circumstances having little to do with the racer's natural endowments, such as the income his or her parents earned, number of siblings, quality of school, and the community's economy. Gender, for one, especially as it interacts with marriage and fertility, is far more consequential than intelligence. Our estimates indicate that a young woman has to score forty-one points higher on the AFQT than a twin brother with identical schooling in order to reduce her risk of poverty to

SOCIAL CHANGE VERSUS IQ CHANGE

Early in *The Bell Curve* its authors acknowledge that much of the social change in recent American history cannot be explained by IQ. The changes are just too large. The great increases in crime in the 1960s and in unwed motherhood in the 1970s, for example, have to be explained in some other way. And yet, by the middle of the book, Herrnstein and Murray suggest that a modest shift in the average American IQ might cause great social change. For example, they calculate, based on their statistical models (which we have challenged here), that a decline of 3 points in the average IQ—a decline they fear differential breeding will bring—would add about 10 percent to the proportion of American children born out of wedlock and to the proportion of American men jailed.

If we accept for the moment the questionable logic of that exercise* and accept their numbers, we can work the algebra backward: How much change in the average American IQ would have been needed to cause the roughly 100 percent increase in the percentage of men in prison during the 1980s? Answer: an IQ drop of about 25 points in ten years. How much change in the average American IQ would have been needed to cause the roughly 150 percent increase in the percentage of children born out of wedlock between 1970 and 1990? Answer: a drop of about 55 points in twenty years (from "very bright" to "very dull" in Herrnstein and Murray's terms). What change in the average American IQ would have been needed to create the roughly 50 percent decline in the overall poverty rate between 1960 and 1978 and then the roughly 25 percent increase in the poverty rate from 1978 to 1992? Answer: about a 25-point rise in the first eighteen years and a 7-point drop in the last fourteen years. And yet, the most changeability in IQ that Herrnstein and Murray can claim is that the average IQ for the nation may move three points in a whole generation.**

Why all the fuss about IQ, then, if by their own calculations IQ changes are so overshadowed by socially caused changes?***

* The exercise is questionable on several grounds. For example, it assumes that a person's *absolute* IQ determines behavior. Yet, Herrnstein and Murray themselves note that "it is one's position in the distribution [i.e., *relative* IQ, which would not change] that has the most significant implications for social and economic life as we know it" (p. 309).

** The statistics are drawn from the *Statistical Abstract 1994*, tables 342, 100, and 727, and the implicit formulas on p. 365 of *The Bell Curve*.

*** Herrnstein and Murray allude to an answer. In discussing crime (p. 239), they admit that IQ can say little about social change, but they contend that some people—i.e., the "dull"—are more affected than others by societywide change. Those with the "greatest tendencies to break laws" would be most affected, and "understanding those tendencies is the business of theories at the psychological level." Fine. Psychology deals with the individuals, and it might highlight who would respond to social change. But then why suggest in *The Bell Curve* that identifying vulnerable individuals says much about change at the *societal* level; why imply that it is change in those *individuals* that drives the changes in society?

his. And for unmarried women, the risk of poverty is very great indeed. In this way, *The Bell Curve*'s most critical empirical claim turns out to be false. Intelligence—even if it were validly measured by the AFQT—does not "dominate" the social environment as a determinant of poverty; quite the reverse.

We can illustrate our conclusion with figure 4.7. The first part of the figure repeats figure 1.2. It shows how much intelligence, as measured by the AFQT, explains the distribution of income in the United States. We see, again, that it explains little. If everyone had the same AFQT scores but still differed from one another in social background, inequality in income would be essentially unchanged. The second part of the figure shows what would happen if, on the other hand, people were all equal on family origins, social environment, and gender but still varied in their AFQT scores.[58] Income inequality would be greatly reduced.

And yet we have not accounted for most of the inequality in income. Perhaps some of the remaining 63 percent of unexplained variation can be accounted for by other, unmeasured attributes of individuals—energy, looks, charm, whatever—or by other, unmeasured attributes of their social situations—grandparents' legacies, social contacts, the industry they work in, and so on—or, as Christopher Jencks has suggested, by simple luck. But much of that remaining inequality can only be understood by leaving the individual level of analysis and looking at the social structure of inequality.

We have, in this chapter, looked at who climbs how high on the ladder of success (particularly looking at those who fall to the bottom), but we have not looked at the construction of the ladder itself. To use another metaphor, we have examined what influences the order Americans finish in the race to success, but have not addressed the prizes racers win. In the next chapter we turn to the prizes—income, wealth, and security—that winners

If all adults had the same test scores (but different family origins and environments), inequality of household incomes would decrease by about 10 percent.

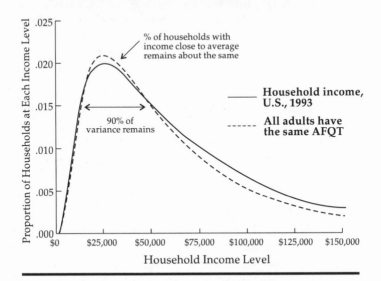

If all adults had the same family origins and environments (but different test scores), inequality of household incomes would decrease by about 37 percent.

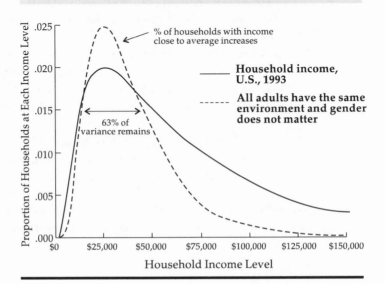

and losers receive. We see how the distribution of prizes has changed over time and how the United States compares with other rich countries in their distribution. We will see, thereby, even more strongly how the real leverage in shaping inequality lies with the choices we, as citizens, make.

On opposite page:

4.7. Explained Variance in Household Income Accounted for by (above) Intelligence Alone versus (below) Social Environment and Gender Alone (*Note*: For method, see text and notes. *Source*: Authors' calculations)

The Rewards of the Game:
Systems of Inequality

I N 1974 the typical corporate chief executive officer in the United States received $35 in pay for every dollar the average worker in manufacturing received. In the 1990s the typical American CEO received over $120 for every average worker's dollar. This change strikingly illustrates how rapidly inequality can change (up to $225 in 1994).[1]

Around 1990 the typical Japanese CEO earned only ¥16 for every yen earned by the average industrial worker, the typical German CEO made DM 21 for every Deutschmark of the average worker, and the typical British CEO received about £25 for every pound earned by the average working person.[2] The contrast between these numbers and 1990's 120:1 ratio in the United States clearly illustrates how much greater inequality in pay is here than it is elsewhere. In addition, executives in other countries pay higher taxes on their relatively lower incomes than do American executives, exemplifying the great difference between us and other nations in after-tax inequality as well.

The CEO numbers point us away from focusing, as we did in chapter 4, on who wins and loses in the race for riches to looking instead at how those races are set up, in other words, at *systems of inequality*. In the previous chapters, we demonstrated that intelligence is not an immutable single trait, is not well-measured in *The Bell Curve*, and does not account for who succeeds or fails in life. In this chapter, we show that, even were *The Bell Curve* correct on all those matters, its psychological reductionism fails to explain the system of inequality. Using history, geography, and economics, we show that social structure and social policy shape inequality.

The Bell Curve addresses the timeless question: Why do some people have more than others—more money, a better standard of living, a higher quality of life? Herrnstein and Murray offer a timeless answer: Because some people have more natural talent than others do. To the more timely question of why some Americans today have more than others do, they answer: Because that is how today's market rewards Americans' natural talents. Whereas once employers paid for strong backs, today they pay for strong minds. In either case, inequality arises from the operation of a free market rewarding people according to their talents. This view of inequality assumes that the unequal distribution of outcomes we see around us essentially results from the unequal distribution of natural skills; the economy confirms that individual inequality.

The previous chapter showed that this "natural inequality" viewpoint misrepresents why individuals end up where they do in the distribution of income. This chapter will show that it fails altogether to explain *systems* of inequality—how much economic inequality there is and what rules determine the kinds of people who get ahead. Even if talent largely determined which individuals won and which lost in the race to get ahead, that would not explain the disparity in the rewards winners and losers get; that is, it would not explain the *degree* of inequality. Even if natural talent explained who became a film star, sports hero, or mortgage banker, it would not explain why people in those positions earn 200 times what child-care workers earn. Nor would natural talent explain why the richest 1 percent of Americans own about 40 percent of all the private wealth in America rather than, say, only 20 percent of that wealth.[3] It would not explain why Americans who lag in the race for financial success still get free primary education for their children but do not get free medical care for their children, or why the also-rans are awarded subsidized old-age pensions but not subsidized housing. To understand the nature of inequality in America, we must look at economic, social, and political conditions—the *systems* of rules and rewards that shape inequality.

An analogy might be instructive. Talent, whatever combination of gifts or hard work created it, partly determines which baseball players become wealthy major leaguers and which forgotten journeymen. But their fortunes are also determined by the number of major and minor league teams, by recruiting procedures, by rules about the draft, call-up options, and free agency, by rules of the game such as the size of the strike zone and the designated hitter, by the market value of franchises and television contracts, by collective bargaining agreements, and so on. Free agency, for example, sharply increased inequality in salaries among major league players.[4] Scarcity is crucial: So few positions are available in professional baseball that many highly talented players never earn a living playing the game. If the rewards of baseball—where pay is so strongly and directly tied to visible and measurable performance—are conditioned so much by circumstances other than individual talent, that must be even more so for the careers of ordinary Americans where the connection between talent and reward is so much harder to assess.

Herrnstein and Murray and their ideological fellows would probably

103

answer that variations among systems of inequality are basically variations in the free market, the difference between, say, a low-tech economy, such as America in our great-grandparents' day, and a high-tech one such as ours. The widening of the gap between rich and poor that has appeared in recent years is, they would argue, essentially the consequence of economic changes that increasingly reward natural, "cognitive" talent. This is a woefully incomplete account.[5]

In this chapter we will show how changeable systems of inequality are. They change in response to economic circumstances, certainly, but also in response to political decisions. Structures of inequality have varied over the course of American history. And most modern societies—technological, capitalist, and wealthy like America—have systems of inequality distinct from our own. We shall show that inequality is a property of how societies are structured, not of how individual talent is distributed. We shall also show that widening inequality in the United States today is not primarily the consequence of natural talent operating in a natural market, but the consequence of political decisions that we, as citizens, have made and pursued through government policies. And we shall show that we have it in our power to alter inequality; we are not constrained by a dismal choice between equality and a high standard of living. Indeed, evidence suggests that the level of inequality we have may be retarding our economy. We can have—have had, as others now have—more growth and more equality.

A Brief History of American Inequality: The Long-Term

Over the nation's history, the inequality that Americans have experienced has varied greatly. Both the rules and the rewards of the "game" have changed. Economic historians can estimate the extent of inequality over a span of two centuries, although the farther back they go the harder it is, of course, to glean precise numbers. Still, the best estimates lead to the following, simplified, history of American inequality.[6]

At the birth of our nation, America was far from a society of equals. On the top was a tiny elite comprised of wealthy merchants and owners of great estates. At the bottom, forming about one-fifth of the non–American Indian population, were African slaves who did not even own themselves. But the bulk of the free population lived in the rough equality of poor yeoman farmers. After about 1800, cities grew, commerce expanded, factories opened, new kinds of white-collar jobs appeared, and entrepreneurial opportunities arose. Americans, many fleeing from their farms, grabbed those opportunities, swelling the numbers of factory laborers, middle-class employees, and small-scale entrepreneurs.[7] In grasping new opportunities, Americans also accentuated the inequalities of income and wealth among them. Immigration magnified that inequality by placing millions of workers at the bottom rung of the job ladder. This increasing division among Americans reversed briefly around the Civil War: The economy boomed and many workers went off to battle, thus making incomes more equal for a while.[8]

During the Civil War two major policies also altered inequality in America. First, Lincoln emancipated the slaves, taking "property" from rich plantation owners and handing it to the "property" itself. The end of the Reconstruction era, however, returned power to the antebellum elites, doomed blacks to a lower-caste position for generations, and set back the trend toward equality. Second, Lincoln's Republican party passed the Homestead Act. This law virtually gave away farmland to Americans who were willing and able to work it. It was manifestly an effort at "social leveling." The Homestead Act and its successor laws were, at best, only partially effective,[9] but they did move white Americans toward slightly more equality.

The earlier trend toward greater *in*equality resumed after the 1860s. A short period of equalization occurred again around World War I, but inequality again widened in the 1920s. The Great Depression and World War II stopped and sharply reversed the tendency toward inequality. The new tendency toward more economic *equality* continued through the 1950s and 1960s, decades of growing prosperity for most Americans. Economists of the day projected slow but continued equalization of income and wealth into the future.[10] They were wrong, as we shall see later.

We can summarize this history: In the nineteenth century, America's economy took off and inequality grew (excepting the emancipation of the slaves); then, in the mid-twentieth century, America's economy matured and inequality narrowed. Between World War II and 1970, middle- and working-class Americans reaped the benefits of economic growth and government programs; they grew richer faster than did rich Americans. By one estimate, the proportion of families in poverty dropped from 56 percent in 1900 to 14 percent in 1972.[11] In political language, the postwar period saw most Americans—with certain groups, particularly blacks, notably excepted—realize the "American Dream."

We should also think of equality not simply in terms of monetary wealth, but in terms of living standards and quality of life.[12] In that sense, twentieth-century equalization was yet more substantial than our account suggests. Virtually all Americans gained basic education and literacy. The vast majority acquired economic security through old-age pensions (mostly social security), unemployment insurance, disability coverage, Medicare, and Medicaid. Most Americans benefited widely from improving public health conditions, and their life expectancy lengthened. (Notably lagging behind were blacks. They were systematically excluded from many support programs, such as social security, until recently.)[13] Almost all Americans found spouses as lifetime spinsterhood and bachelorhood became rare, both a consequence and a cause of increased security.[14] Most Americans—whether winners or also-rans in the race to get ahead—shared in the good life, making America circa 1970 as equal a society as it had ever been.

As we have noted, it is difficult to measure inequality precisely before recent decades. But the lines in figure 5.1 summarize what historians have been able to learn. The vertical axis represents the degree of inequality. The dashed line displays changes in inequality of wealth and income; the solid one includes other qualities of life, too, such as public health and education. One conclusion we draw is that the magnitude of inequality has varied greatly in the course of American history.

This brief review of American history also suggests some conclusions about the sources of inequality. It would be hard to explain these trends in terms of intelligence, whether rising, falling, stretching, or contracting. It would also be hard to explain it in terms of market demand for high-IQ workers. That is how Herrnstein and Murray try to explain why inequality increased *after* 1970: New high-skill, high-stakes, high-pay jobs appeared and rewarded the people who had high IQs. One problem with their explanation is that the American economy changed in just these ways *before* 1970, too. Between 1930 and 1970, the percentage of Americans working on farms dropped from 21 to 4 percent, those in blue-collar jobs shrank a bit from 40 to 35 percent, and the percentage holding white-collar and professional jobs grew substantially, from 29 to 48 percent of all jobs.[15] Such an increase in the demand for brainpower and book-learning should have, if we believe *The Bell Curve*, increased inequality between 1930 and 1970, but Americans actually became *more equal* in those years.

The reasons historians give for twentieth-century equalization include greater economic productivity, unionization, democratization of higher education, and government intervention. Government intervention, which we shall discuss in more detail in chapter 6, includes social security, mini-

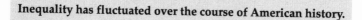

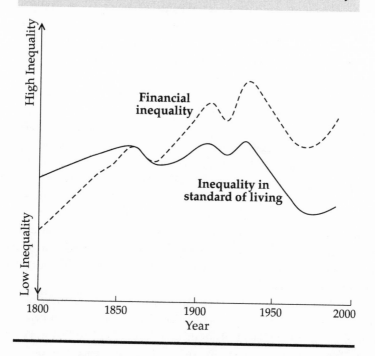

5.1. Estimates of Inequality from 1800 to Today (*Note*: Lines are authors' estimates of trends described in sources listed in note 6)

mum-wage legislation, expansion of public universities, the GI Bill, housing subsidies, and many other programs that effectively brought economic growth and middle-class life to more Americans.

INEQUALITY SINCE THE 1970s

During the 1980s politically engaged academics argued over whether the American middle class was "disappearing" and the nation was splitting into rich and poor. Defenders of freer markets denied that this was happening, blaming misinterpreted data or the business cycle.[16] But the answer is clear now. The trend toward a more equal society that developed in the twentieth century stopped in the early 1970s. Income divisions widened and continued expanding through the 1980s.

A Case of Historical Amnesia: Caring for the Needy

While *The Bell Curve*'s treatment of American history is generally problematic, one of its greatest historical errors is the discussion of how a past
America dealt with the unfortunate.* The authors of *The Bell Curve* recall
a time when neighbors provided all the "safety net" that the poor needed.
But such memories are notoriously unreliable.

There really was no such time in America. If the unfortunate were long-
time residents of the town, victims of bad luck, and considered morally
upright, then neighbors might pass the hat, allot funds from the municipal
treasury, or board the destitute with a local family. Even so, the support
was typically small and grudging. If the needy did not pass these tests,
especially if they were newcomers, then usually nothing was forthcoming.
(One historian wrote of a colonial Puritan village that "poor persons were
aided if they were members of a townsman's family, otherwise they were
sent packing no matter how hungry they might be.")** From colonial
days, when towns "warned out" newcomers whom residents feared might
become public dependents, to the Great Depression, communities' major
response was to move the needy out.

(We have forgotten how many Americans were on the move. Americans used to be much more mobile than they are now. Upon being laid
off or losing a farm, hundreds of thousands moved on to find work for a
while and then move again. These people do not appear in our Andy
Hardy–like memories of small-town America, because, being poor and
transient, they were anonymous residents on the wrong side of the tracks.
But their stories belie the nostalgic image of the small town that succored
the needy.)

The local structure for helping the needy, pinched as it was, satisfied
most American voters well enough for generations. But the Great Depression destroyed that structure. Many long-time, middle-class, and "worthy"
residents also became destitute. Would-be charity-givers themselves became needy. Local philanthropies, even where supplemented by town and
county government contributions, could not keep up. In response, New
Deal programs funnelled massive amounts of money into local communities directly and through work projects. Whether or not these programs
actually ended the Depression, federal intervention certainly supported

* See, especially, pp. 536–40.
** Lockridge, *A New England Town*, p. 15.

many of the needy through the hard times.*** By now, the federal role in sustaining the poor, sick, and elderly has become essential, even for private charities.****

Some have proposed that America today deal with the destitute by returning to private philanthropies and local communities the responsibility to do so, returning to an earlier "safety net." Advocates of such policies can appeal to nostalgia but they cannot appeal to history. The history of the local community shows that its "safety net" was composed mostly of holes.

*** For general overviews, see Katz, *In the Shadow of the Poor House*; Trattner, *From Poor Law to Welfare State*; Keyssar, *Out of Work*; and Sautter, *Three Cheers for the Unemployed*. For the history of the New Deal's role, see also Chambers, *Seedtime of Reform*; and Brock, *Welfare, Democracy, and the New Deal*. For monographic studies of the needy in earlier eras, see, for example, Althschuler and Saltzgaber, "The Limits of Responsibility"; Vandal, "The Nineteenth-Century Municipal Responses to the Problem of Poverty"; and Monkkonen (ed.), *Walking to Work*.

**** Estimates are that replacing a proposed $400 billion cutback in federal programs for the needy by the year 2002 would require immense and improbable increases in private giving to charity (Steinfels, "As Government Aid Evaporates").

In the first roughly twenty-five years after World War II, American families of all classes shared in economic growth; since 1970 only the richest families have seen a significant rise in their standard of living. We saw in figure 1.1 the trend since 1959. In the ten years between 1959 and 1969, incomes for the richest 20 percent of households grew from $29,000 to $36,000 per person (in real spending power); incomes for the middle 60 percent of households grew from $11,500 to $16,000 per person; incomes for the poorest 20 percent grew from $2,900 to $5,400 per person. The percentage increase for the poorest households was larger, at 6.5 percent growth, than the others. In the twenty years after 1969, the incomes of the top 20 percent have risen an additional $28,000 per person, the incomes of the middle 60 percent have gone up by a modest $4,600 per person, and the incomes of the poorest 20 percent of households have actually fallen by $200 per person. For several reasons discussed below, the pattern displayed in figure 1.1 may underestimate the declining conditions of the middle group.

These converging and then diverging fortunes are further indicated in the data on the shares of the nation's total income that went to different

From 1930 to 1970, the lowest–income 40% of American households began receiving almost as much of American income as the highest–income 5%, but that trend reversed after 1970.

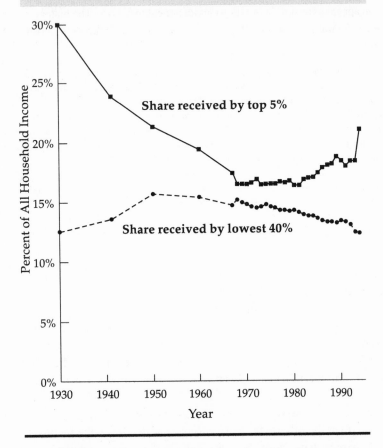

5.2. Percentage of All Household Income Received by Highest-Income 5 Percent and Lowest-Income 40 Percent of Households, 1930–1994 (*Sources*: U.S. Bureau of the Census, *Historical Statistics*, p. 301; Ryscavage, "A Surge in Growing Income Inequality?"; U.S. Bureau of the Census, "Income and Poverty: 1994, Highlights")

income groups. Figure 5.2 shows that, in 1930, the highest-income 5 percent of American households combined received over twice the total income that the bottom 40 percent of American households combined received. The highest-income 5 of every 100 families received about one-third of all the income received by American households that year, while the lowest-income 40 households out of 100 received one-eighth of national income. By 1968 equalization had developed to the point that the bottom two-fifths finally brought home almost as much as the top one-twentieth (15.3 percent versus 16.6 percent). Households between the 40th and 5th percentiles had increased their share from 58 to 68 percent (not shown). After the 1960s, however, the income share of the top 5 percent rose sharply and that of the lower 40 percent of families dropped sharply, reversing the earlier trend toward equality. The unequalizing trend has continued into the 1990s.[17]

Economists and sociologists, working from different perspectives, have analyzed and reanalyzed the available data and have come to a consensus.[18] Inequality in income expanded greatly after about 1970. People at the top made *more* money and people in the bottom half made *less*. In recent years, the chances of low-income Americans moving into the middle class have dropped and the chances of middle-class Americans moving into poverty have grown.[19]

The growing inequality in annual income displayed in the figure actually understates the magnitude of the change. One reason is that gaps in annual income accumulate as some families invest and others pile up debts. Wealth has become even more unequal than has annual income. From about 1975 to 1992, the *wealthiest 1 percent* of families' share of the national household wealth rose from about 22 percent to about 42 percent; in the 1980s the wealth held by the *poorest 40 percent* of families actually dropped in absolute value. The middle class has also felt the strain. In the 1960s the middle one-third of Americans saved about 5 percent of their annual income; in the 1980s they saved virtually nothing.[20]

Another reason the annual income changes shown in figure 5.2 understate the change is that the increase in inequality has been most acute in the younger generations. The proportion of young men earning a "family wage"—that is, enough income to keep a family of four out of poverty—has fallen sharply since the 1970s. Figure 5.3 shows the proportion of full-time, year-round employed men, divided into a few types, who earned at least a family wage from 1964 to 1994. We can see that the proportion that earned enough went up substantially from 1964 to 1974. But then the proportion declined. It declined dramatically for high school dropouts and for

The proportion of full–time employed men whose earnings could keep a family out of poverty rose until 1974 and then dropped.

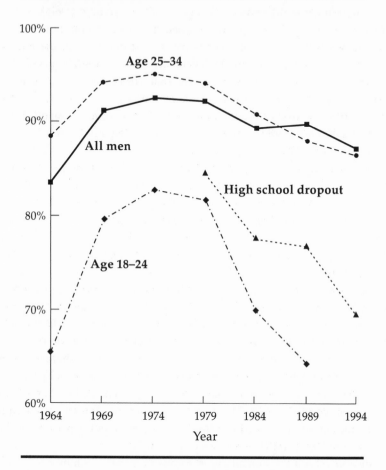

5.3. Percentage of Full-Time Workers Who Earned Enough to Keep a Family of Four Out of Poverty, 1964–1994 (*Note*: The measure is the percentage of full-time workers whose annual earnings exceed the poverty line for a family of four. *Sources*: U.S. Bureau of the Census, "Workers with Low Earnings"; and Jack McNeil, U.S. Bureau of the Census, unpublished tables, November 1995)

18-to-24-year olds, but it even declined for workers aged 25 to 34.[21] Similarly, men turning 30 around 1990 were notably less likely to have already attained a middle-class income than were men who had turned 30 around 1980.[22] These figures underestimate the sinking fortunes of workers because they count only fully employed men; in recent years, fewer men have been fully employed.[23] Less able to save, less able to move up a career ladder, younger men are seeing differences in lifetime earnings and wealth widening yet further. All these compounding changes mean that inequality in economic security and independence has accelerated sharply. The proportion of Americans, aside from the elderly, who are poor has increased. Unlike the leveling that accompanied the economic growth of the 1950s, the economic growth of the late 1980s failed to stop or seriously slow this unequalizing trend.

Although young workers have been the most vulnerable to the economic dislocations since 1973, older male workers have also been affected. This can be seen most clearly in figure 5.4. It shows what has happened to the annual income (cost-adjusted) of men as they grew older, contrasting the 1950s and 1960s with the 1970s and 1980s. The gray arrows show what happened to men who started each decade aged between 25 and 35—the income they started the decade with and the income they ended the decade with, now aged 35–45. The black arrows show that pattern for men who started each decade aged between 35 and 45. In the period before 1973, men could expect to earn more, on average, at the end of the decade than they did at the beginning. For example, men who were between 25 and 35 in 1950 made about $12,000 in constant dollars, and by 1960 they averaged about $17,000. However, from 1973 to 1983, men who had started the decade aged 35–45 (black arrow) saw drops in income; men who had started the decade aged 25–35 (gray arrow) had stagnant incomes. And from 1983 to 1993, men in the 35–45 cohort had flat incomes. Men in the 25–35 cohort had growing incomes, but that cohort of 30-ish men started so far down that even the gains they had made by 1993 left them behind men who had been the same age in 1973. The assumption that men will move up the salary ladder as they mature now seems untenable.

M.I.T. economist Frank Levy, who developed the original version of figure 5.4, offers the image of "yesterday's $25,000 steelworker who now clerks in a K-Mart at $4.25 an hour."[24] But it is not only that workers in declining industries like steel had to move on to new jobs. Men in many sectors of the economy were displaced. A 1995 Census Bureau report compared job mobility in the late 1980s to mobility in the early 1990s and found that workers in the 1990s were more likely to lose jobs. And full-

Before the 1970s, 30– and 40–year–old men made major gains in income over a decade, but that has not been true since 1970.

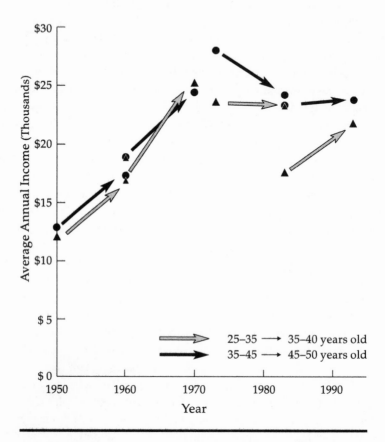

5.4. Income Changes Over a Decade for Men in Their Thirties and Forties, 1950–1993 (*Note*: Modeled on Levy, *Dollars and Dreams*, p. 81. *Sources*: U.S. Bureau of the Census, *Current Population Reports*, various series; and *Statistical Abstract 1994*)

time workers who lost their jobs more frequently ended up as part-time workers; even those who found a full-time job after a period of joblessness saw a 20 percent drop in their average weekly earnings.[25]

There has been some controversy about the claim of growing inequality that we just made. Few observers deny that the rich have gotten much richer faster than everyone else in the United States. But the dissenting voices claim that the economic situations of average and poor Americans have also improved in the last decade or so, if not as rapidly as those of the wealthy. If that were true, one might be able to claim that the fortunes of the wealthy have trickled down to the rest of Americans.

Some argue, for example, that the way the Bureau of Labor Statistics calculates the cost of living exaggerates the year-to-year cost increases, perhaps by as much as 1.5 points. All the numbers we have used here, which adjust incomes for increases in cost of living, would therefore underestimate the growth in real earning power and overestimate the growth in poverty. Instead of concluding that the median earnings of fully employed men dropped 12 percent in buying power from 1979 to 1994, we would conclude that they rose 14 percent. Recalculating the price index would not change our conclusion that the *gap* between classes has increased, "but instead of the usual story of the rich getting richer and the poor getting poorer, the new story would be that the rich got a lot richer while the poor held their own."[26] The debate over this claim can get arcane, dealing with exactly how to weigh apples and oranges in the typical consumers' "market basket" and how to account for changes in products. Is, for example, a 1995 television set that costs the same as a 1975 television set but has stereo sound and a cable socket a "cheaper" television set? Is a visit to an HMO doctor the same "value" as a visit to a private practitioner? A few points seem clear, however: One is that the debate over earning power applies only to those who are earning. As noted earlier, fewer men are earning incomes today than before.[27] A second is that the "big-ticket" items of the middle-class life-style, such as a home and college education for the children, have risen in "real" cost much more rapidly than the little items that fill up the statisticians' market basket. Americans have had to work a lot harder and go into more debt to attain those pieces of the "American Dream."[28] A third point is that between 1975 and 1993 many more wives with children at home started working, from fewer than half to over two-thirds of them[29]—a great many because they believed, rightly or wrongly, that maintaining a middle-class life-style required it. As mothers took paying jobs, much of the free labor they had contributed to the household in cleaning, cooking, child care, and the like had to be purchased or done without. If one takes these points into account, the decline in living standards has probably been *greater*, not less, than the crude numbers indicate.

Another line of defense charges that the apparent increase in poverty among the nonelderly is also a statistical anomaly. Some commentators point out that often government aid to the poor such as food stamps is not counted as income. But add into the calculations government assistance and the trends are still the same. (Recall that the problem of inequality is not so much a problem of the poor, although it is that, too, as it is a problem of *most* American families falling behind the wealthy, as well as behind their own expectations.) Others have argued that the increase in poverty is the product of family change—the increase in divorce and in the number of women bearing children out of wedlock. This argument implies that the structural situation is fine, the problem is that many women have, in effect, *chosen* to be poor by making poor life decisions. This argument also fails. First, divorce and single-parenthood are often (if not usually) the results of economic strain, typically the inability of a man to help support his family. Therefore, much of the family dislocation itself is the consequence, not the cause, of economic dislocation. (Recall that we found in chapter 4 that having been poor was the key determinant of whether a woman would later have a child out of wedlock.) Second, even if we were to assume that every divorce and out-of-wedlock birth were not the result of economic distress, such family changes account for only about one-third of the rise in inequality. And, at the same time, other family changes have actually slowed down the inequality trend. More lower-income women are working than before, and they are having fewer children. If not for these last two changes, the trend toward greater inequality would have been more severe still. In sum, the growing inequality is real, it is substantial, and it is not simply a matter of the rich outpacing an advancing field; the rich have been advancing and the rest have been retreating.[30]

The post-1970 increase in income inequality largely arose from increasing inequality in workers' *earnings*. (Increasing *wealth* inequality was, in great measure, the result of the escalation in the value of financial instruments relative to owner-occupied homes during this period.)[31] Earnings, especially those of younger men, have become much more unequal. In particular, earning differences by education have widened. During the 1980s the hourly wage (adjusted for inflation) went up 13 percent for men who had graduated college but declined 8 percent for men who had dropped out of college, dropped 13 percent for men who had only graduated from high school, and plummeted 18 percent for male high school dropouts.[32] Inequality in earnings has also expanded among workers of comparable education, and workers in all sorts of industries have been affected. In turn, explaining why wages have diverged is more difficult.

University of Massachusetts economist Barry Bluestone lists ten different theories for explaining why the market seems to be rewarding education and skill more than it used to.[33] One part of the explanation appears to be technological change, especially computerization, which would make more highly educated workers more valuable to employers. Some evidence points to this as a factor, although it is still a controversial claim.[34] Many technological changes simplify jobs instead of making them more complex and therefore encourage employers to hire less-skilled rather than more-skilled workers.[35] "Deindustrialization"—the process by which manufacturing jobs move overseas or are automated, leaving more poorly paid service jobs in America—explains what happened in terms of reduced demand for the work of the less-educated. Generally, competition from other nations' workers has depressed middle Americans' wages. When, for example, software programming can be done in India at a fraction of the cost here, it is hard for American workers to demand higher pay.[36] (This pressure seems not, however, to have depressed CEO salaries, as we saw earlier. Indeed, Disney executive Michael Eisner banked a record $203 million the year his company suffered a 63 percent drop in profits.)[37] Also contributing to the free-fall in less-skilled workers' wages are the weakening of unions, the stagnation in the minimum wage, and cutbacks in higher education (see chapter 6).

It is important to understand that this widening inequality in earnings has occurred, not only in the United States, but in most affluent Western nations during recent years. However, as we shall see in the section below that compares the American experience to that of other nations, expanding inequality has been greater and its consequences more severe here than anywhere else (except perhaps for the United Kingdom).[38]

Analysts often treat inequality in earnings as simply the result of "market" operations and separate those effects from the effects of governments' after-market interventions—taxes and transfer payments. We will do that, too. But we first note that the distinction between market and policy is misleading. The market is permeated with policy.[39] (See box, p. 118.)

By its taxes and by its "transfers"—social security, Medicare, food stamps, unemployment insurance, and the like—government can blunt inequalities in income created by inequalities in earnings from the market. American government today does reduce inequalities of income, but largely for older people.

How the "Free" Market Rests on Government Policy

What people earn in the labor market cannot be separated from policies that structure the market. We discuss such policies in detail in the next chapter, but pause here to consider just a few examples of how government policies shape difference in earnings:

—Licensing laws: Governments stipulate the requirements necessary to practice a wide range of professions, from hair-cutting to neurosurgery. The tighter those requirements, the fewer the practitioners, and the higher their earnings. Imagine what would happen to the incomes of the top-earning 5 percent of Americans if entry into medicine, or law, or the professorate were made considerably easier? (As professors, we are not necessarily recommending this move.)

—Direct subsidies: Direct subsidies, such as agricultural farm supports and urban transit construction, divert income from some people to others. American economic development in the early nineteenth century depended heavily on federal gifts of land to the states and on state borrowing for infrastructure. (Jeffersonian-Jacksonian Democrats largely opposed such spending, and Hamiltonian Federalists, predecessors of today's "Wall Street" Republicans, supported it.) Later, sizable subsidies to railroad companies spurred the interconnection of American towns.

—Laws governing property and finance: Fundamental to the economic system are the laws that govern property. In the early nineteenth century, American courts provided critical rulings that enabled the creation of limited-liability corporations and that protected businesses from paying for the incidental damages they caused. Different rulings would have meant different earnings. Later, the Supreme Court ruled that the Fourteenth Amendment's protections for people extended as well to corporations, further aiding the expansion of the large corporate sector.

Other examples of how policy shapes the market, such as unionization rules, tax deductions, and regulations, will be discussed in chapter 6. But even these few illustrations show us that the way the "market" apportions income is far from a pristinely economic process. It is embedded in a set of policies, many of which may be so longstanding that we assume them to be "natural." Nevertheless, they are policy *choices*. "There is no such thing as a free lunch," the laissez-faire economists remind us. True, and there is also no such thing as a "free" market.

Since 1966, the poverty rate among the elderly has dropped by half; it has increased among other Americans.

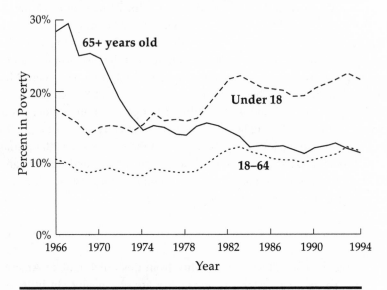

5.5. Rates of Poverty by Age Group, 1966–1994 (*Sources*: U.S. Bureau of the Census, *Current Population Reports*, series P-60, no. 178, and series P-23, no. 188; and "Income and Poverty:1994, Highlights")

Not so long ago, older people were poorer than the rest of Americans, and their poverty was a social problem of wide concern. But, as we first pointed out in chapter 4, poverty among older people has dropped dramatically (see figure 5.5). By 1994 under 12 percent of older people were poor, compared with 15 percent of the nonelderly. The single major reason for this reversal is social security. For decades, it has allowed older people to retire early. In 1940, 42 percent of older men worked; in 1994, only 17 percent did. Then, in more recent years, cost-of-living adjustments accelerated the increase in social security payments. In the 1980s, while the median income of all American households increased just 5 percent, the median income of older households increased 20 percent.[40] By 1993 the *wealth* of the typical older household was over twice the national average.[41] America's social policy has successfully fought the "war on poverty"—but mostly on behalf of older people.

Inequality has widened since 1970 not just in money, but also in the quality of life money can buy. We see it, for example, in homeownership, something Americans value greatly. Between 1983 and 1994 rates of homeownership fell for every age group under 60 while rising for every age group over 60. For example, the percentage of heads of household 40–44 years old who owned their homes dropped from 73 percent to 68 percent, but the percentage of those aged 70–74 who owned their homes grew from 75 percent to 80 percent.[42] People's sense of security relies in part on having health insurance, to take another example, but the proportion of Americans covered by any health insurance (increasingly this has meant Medicaid) for an entire year has fallen recently, from about 87 percent in 1987 to 85 percent in 1993 and is still falling (though not among older people, covered as they are by Medicare).[43] More dramatically yet, about one-third of men who changed jobs recently found that they had lost health insurance in the process.[44] Also paralleling the widening income and wealth gaps is a widening gap in health and mortality among social classes.[45] Inequality defined broadly in such ways is also growing.

Conclusion

What have we learned about inequality from this quick look at American history? We see that systems of inequality are changeable. In little more than a generation, Americans became notably more equal and then notably more unequal. These changes cannot be explained by changes in individuals' "natural" talents, be it IQ or other inherent traits. (Changes in Americans' *acquired* traits probably played a role. Far more Americans were college graduates in the 1980s than in the 1940s.) Even if, as some testing data suggest, Americans became "smarter" during the twentieth century (see chapter 2), inequality changed in different directions, and the scale of the changes was just too great. Market forces also cannot explain these variations in inequality. Certainly, technological developments and global trade have helped shape inequality in the United States, but much has been under the control of policy—policy that structures how the market works and policy that corrects how the market works. This point emerges again when we compare the United States with other nations.

INEQUALITY HERE AND THERE

A glance behind us to American history shows that our pattern of inequality is far from fixed or naturally determined. A glance sideways to other wealthy nations makes the same point. The United States has the greatest degree of economic inequality of any developed country. It is a level of inequality that is not fated by Americans' talents nor necessitated by economic conditions but is the result of policy choices. The nations with which we will compare the United States are also modern, affluent, democratic, and capitalist—they are our competitors in the global market—and yet they have ways to reduce inequality and remain competitive.

At the beginning of this chapter, we noted how much wider the gap in earnings is between CEOs and average workers in the United States than it is elsewhere. America's distinctive inequality is greater even than these illustrative numbers indicate. In general, our high earners earn relatively more and our low earners earn relatively less than do workers elsewhere. The best and latest evidence on how nations compare in levels of inequality comes from the Luxembourg Income Study (so named because the project is headquartered in Luxembourg). Social scientists affiliated with the study have collected detailed, comparable data on earnings and income from over a dozen nations. Our first use of their research appears in figure 5.6, which speaks to the question of inequality in *earnings*, specifically earnings of men, aged 25–54, who worked full-time, all year during the mid- to late 1980s. (Comparable data on earnings were available for only five nations. We are looking just at men here, because the situation of women in the labor force was in such flux and varied so much among nations.)[46] The vertical line in the figure serves as an anchor for looking at inequality in each nation. It represents the earnings of the average (median) worker. The horizontal bars to the left of the median line display the ratio of the earnings that men near, but not at the top of, the earnings ladder received—those at the 90th percentile in earnings—to the earnings of men at the median. In 1986 the 90th percentile American male worker earned 1.8 times what the median worker earned. The bars stretching to the right represent the same comparison between the median earner and a low-paid worker, one at the 10th percentile of earnings. In the United States, the median worker brought home 2.8 times the amount the 10th percentile worker did. The left-hand bars, therefore, display inequality of earnings between the high-earners and the average; the right-hand bars display inequality between the average and the low-earners. Together, they display total inequality. In the United States, the 90th percentile worker earned five times that of the 10th percentile worker.

These numbers are highest in the United States. That is, the gap in earnings between the rich and the average worker is greater here than elsewhere, as is the gap between the average and the low-paid worker. The contrast between the United States and Europe sharpens further when non-

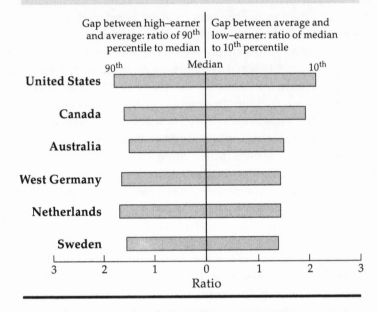

The gap between the highest– and the average–earning men was widest in the United States –– as was the gap between the average– and the lowest–earning men.

Gap between high–earner and average: ratio of 90th percentile to median	Gap between average and low–earner: ratio of median to 10th percentile

5.6. Ratios of Earnings for High-, Median-, and Low-Earners in Six Nations (*Source*: Adapted from Gottschalk and Smeeding, "Crossnational Comparisons," table 1)

monetary compensation is added to the picture. In most European nations, national law requires that virtually all workers have the kinds of benefits such as strong job security and four-week vacations that in the United States only workers with seniority in major firms have.[47]

These national differences expanded in the 1980s, when inequality increased globally. International economic forces widened the gaps between what the better- and the worse-educated earned in most industrialized nations, but this chasm opened up farthest and fastest in the United States and the United Kingdom. (These were the years of Thatcherite reforms that reduced the role of government in the United Kingdom.) Elsewhere, the gap in earnings between the better- and worse-educated widened less, barely at all, or even narrowed. There seems no clear connection between these differences and other economic trends such as growth rates. The reasons lie in government policies, notably the relative power of unions and the expansions of higher education in the other Western countries.[48]

The biggest contrast in income inequality between the United States and the rest of the developed world, however, appears *after* taking into account how government deals with the results of the market. That means accounting for taxes, tax deductions, transfer payments, housing subsidies, and the like. (Again, we note that this before- and after-government distinction underestimates the role of government. Where, for example, governments require employers to provide certain benefits, there is more market equality.)

To look at international differences in *household income*, we turn again to the Luxembourg Income Study. Peter Gottschalk and Timothy Smeeding compiled comparable data on households' disposable incomes—income after taxes and government support, adjusted for household size—in seventeen nations.[49] In Figure 5.7, we use just the figures for nations with over ten million residents in 1980; our conclusions about the United States would be virtually the same if we showed the smaller nations, too.[50] As in figure 5.6, the bars to the left of the median display the ratio of a rich household's income (at the 90th percentile) to an average one's income, while the right-hand bars show the ratio of an average household's income to that of a poor one (10th percentile).[51] The rich-to-average ratio is greatest in the United States, 2.1, as is the average-to-poor ratio, 2.9, and so the rich-to-poor ratio, 5.9 (not shown) is much higher than that of the next most unequal nations (4.0 for Italy, Canada, and Australia).[52] In short, the United States has the greatest degree of income inequality in the West whether one focuses on the gap between the poor and the middle or the gap between the middle and the rich. Even these numbers underestimate America's distinctiveness, because they do not count the sorts of "in-kind" help that middle- and lower-income families receive in most other nations, such as free health care, child care, and subsidized housing and transportation. They also underestimate inequality in America by not displaying the concentration of income at the very top of the income ladder.

Western nations generally take two routes to reducing inequality. As we discuss in chapter 6, some intervene in the market to ensure relatively equal distributions of *earnings* by, for example, brokering nationwide wage agreements, assisting unions, or providing free child care. Others use taxes and government benefits to reduce inequality of *income* after the market. A few do both seriously, such as the Scandinavian countries. The United States does the least of either. If one sets aside older people, who benefit a

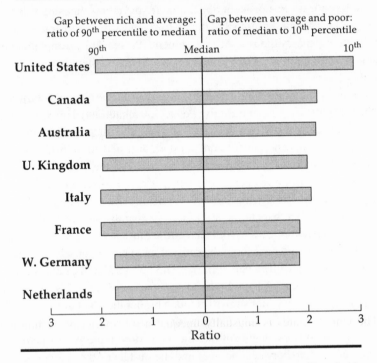

The income gap between the richest and the average household and the gap between the average household and the poorest are both wider in the United States than elsewhere.

Gap between rich and average: ratio of 90th percentile to median | Gap between average and poor: ratio of median to 10th percentile

5.7. Ratios of Incomes for High-, Median-, and Low-Income Households in Eight Nations (*Source*: Adapted from Gottschalk and Smeeding, "Crossnational Comparisons," table 3)

great deal from government action in the United States, the net effect of taxes and transfers here is to leave the degree of inequality virtually unchanged from the way it was determined by market earnings.[53] (See also chapter 6.)

When everything is accounted for, the Western nation with the most income inequality is the United States. But the United States is also exceptionally unequal in terms of *wealth*. At the end of the 1980s, the richest 1 percent of families owned about 40 percent of household wealth here, more than in any other advanced nation; the richest 1 percent owned only 25 percent of the wealth in Canada and 18 percent of the wealth in Great Britain, for example.[54] Add the less tangible features of "wealth," such as vacations and security of medical care, and the conclusion is reinforced that Americans are remarkably unequal.

(Some critics of crossnational comparisons contend that one ought not to contrast the United States to other nations, because the United States is distinct in certain ways. We have so many single-parent families, for example. But even looking only at two-parent families, the United States is still unusually unequal.[55] America also seems exceptionally diverse racially and ethnically. But other Western nations also have ethnic diversity, if not the racial caste system we do. And poverty among American whites *only* still exceeds that among white or majority populations elsewhere.[56] Such reservations do not challenge the conclusion that the United States is unusually unequal.)

America's level of inequality is by design. It is not given by nature, nor by the distribution of its people's talents, nor by the demands of a "natural" market. Other Western nations face the same global competition that we do and are about as affluent as we are and yet have managed to develop patterns of inequality less divisive than ours. Ironically, it was not so long ago that Americans were proud of comparing their relatively egalitarian society to the class-riven, hierarchical, decadent societies of Europe. In the last couple of decades, America has become the more class-riven and hierarchical society.[57]

The United States is unusually unequal and Americans are unusually supportive of this inequality. Surveys show that Americans back moves toward expanding *opportunity* but oppose moves toward equalizing *outcomes*. They endorse wage differences among jobs that are pretty similar to the wage differences that they believe exist today (although the real differences are greater than Americans imagine), and they do not approve of government programs to narrow those differences. In a survey of people in six nations, only 28 percent of Americans agreed that government should reduce income differences. The next lowest percentage was 42 percent (Australians), while in the other countries majorities supported reducing income differences.[58] Whether we have as much opportunity as Americans want is debatable (see chapters 4 and 8), but we seem to have a rough match between the desired and the perceived level of outcome inequality. That may be because Americans think that considerable inequality is needed for stimulating productivity and a high standard of living. Is it?

125

Is Inequality the Price of Growth?

Some commentators straightforwardly defend our current level of in-equality. A congressional report in 1995 conceded that the recent trends toward inequality were real but argued, "All societies have unequal wealth and income dispersion, and there is no positive basis for criticizing any degree of market determined [sic] inequality."[59] Disparities in income and wealth, some analysts argue, encourage hard work and saving. The rich, in particular, can invest their capital in production and thus create jobs for all.[60] This was the argument of "supply-side" economics in the 1980s, that rewarding the wealthy—for example, by reducing income taxes on returns from their investments—would stimulate growth to the benefit of all. The 1980s did not work out that way, as we have seen, but the theory is still influential. We *could* force more equal outcomes, these analysts say, but doing so would reduce living standards for all Americans.

Must we have so much inequality for overall growth? The latest eco-nomic research concludes *not*; it even suggests that inequality may *retard* economic growth. In a detailed statistical analysis, economists Torsten Persson and Guido Tabellini reported finding that, historically, societies that had more inequality of earnings tended to have lower, not higher, sub-sequent economic growth. Replications by other scholars substantiated the finding: More unequal nations grew less quickly than did more equal soci-eties.[61] That fits more casual observations as well: We saw that, in the United States, our era of greatest recent growth was also an era of greater equalization.[62] And we saw, at the beginning of this chapter, that Amer-ica's economic rivals do not need to pay their CEOs exorbitant salaries to give us stiff competition. In fact, during the 1970s and 1980s, America's national wealth did not grow as fast as that of the more egalitarian Euro-pean nations.[63]

Close examination of detailed policies also suggests that greater equality helps, or at least does not harm, productivity. Researchers affiliated with the National Bureau of Economic Research closely examined the effects on economic flexibility (that is, the ability to shift resources to more produc-tive uses) of several redistributive policies used by Western nations—job security laws, homeowner subsidies, health plans, public child care, and so on. They found that such programs did *not* inhibit the functioning of those economies.[64] Indeed, a study of over one hundred U.S. businesses found that the smaller the wage gap between managers and workers, the higher the business's product quality.[65]

This recent research has not demonstrated precisely how greater equality helps economic growth,[66] but we can consider a few possibilities. Increasing resources for those of lower income might, by raising health, educational attainment, and hope, increase people's abilities to be productive and entrepreneurial. Reducing the income of those at the top might reduce unproductive and speculative spending. Take, as a concrete example, the way American corporations are run compared with German and Japanese ones. The American companies are run by largely autonomous managers whose main responsibility is to return short-term profits and high stock prices to shareholders and—because they are often paid in stock options—to themselves as well. Japanese and German managers are more like top employees whose goals largely focus on keeping the company a thriving enterprise. The latter is more conducive to reinvesting profits and thus to long-term growth.[67] Whatever the mechanisms may be, inequality appears to undermine growth. Americans certainly need not feel that they must accept the high levels of inequality we currently endure in order to have a robust economy.

A related concern for Americans is whether "leveling" stifles the drive to get ahead. Americans prefer to encourage Horatio Alger striving and to provide opportunities for everyone. Lincoln once said "that some would be rich shows that others may become rich."[68] Many, if not most, Americans believe that inequality is needed to encourage people to work hard.[69] But, if so, *how much* inequality is needed?

For decades, sociologists have been comparing the patterns of social mobility across societies, asking: In which countries are people most likely to overcome the disadvantages of birth and move up the ladder? In particular, does more or less equality encourage such an "open" society? The answer is that Western societies vary little in the degree to which children's economic successes are constrained by their parents' class positions. America, the most unequal Western society, has somewhat more fluid intergenerational mobility than do other nations, but so does Sweden, the most equal Western society.[70] There is no case for encouraging inequality in this evidence, either.

In sum, the assumption that considerable inequality is needed for, or even encourages, economic growth appears to be false. We do not need to make a morally wrenching choice between more affluence and more equality; we can have both. But even if such a choice were necessary, both sides of the debate, the "altruists" who favor intervention for equalizing and the supposed "realists" who resist it, agree that inequality can be shaped by policy decisions: wittingly or unwittingly, we choose our level of inequality.

CONCLUSION

Either we Americans have come to desire the inequality we live with or the inequality we live with reflects our desires (or both). Certainly, there are values and beliefs—individualism, capitalism, freedom—that can justify creating a more *un*equal society. But many Americans also believe that such levels of inequality are inevitable, the result of inequality in natural ability or of the market or both. This belief has no basis in evidence. Income and, more generally, wealth, standard of living, and quality of life have been more equal in other times and are more equal in other places, without any evidence that talents were more equal earlier or are more equal elsewhere, or that market pressures are different there.

Perhaps today's trend toward inequality will reverse; it has reversed before in American history, as economic and political conditions changed. But the keys to understanding inequality will remain. The degree of inequality we live with is not a "natural" result of either inherent human talents or a "free" market. Certainly, people's skills and societies' economic conditions strongly influence the shape of inequality. But a people, acting through its government, can contract or expand that inequality. The policy changes enacted by the 1995–96 Congress will, certainly in the short run and most probably in the long run, widen inequality. We explore the choices we have in more detail in the next chapter.

How Unequal?

America's Invisible Policy Choices

Americans can significantly alter how much inequality there is among them. In chapter 5, we showed how inequality changes over time and how much it varies from nation to nation. Such fluidity results in large measure from changes and variations in *policy*. In this chapter we focus on several specific American policy choices that shape inequality.

Obvious redistributive programs, such as welfare spending, are not the only policies, or even the most important ones, that affect inequality. Many "invisible" practices are more significant. For example, American housing and road-building programs have largely subsidized the expansion of sub-urban homeownership for the middle class. Other largely unnoticed policies set the ground rules for the competition to get ahead. Just as in base-ball, where the height of the pitcher's mound affects whether pitchers or batters have the advantage (see box, p. 130), so in the marketplace laws and regulations favor some competitors and disadvantage others. We saw in the last chapter that the United States has the greatest inequality in earnings among full-time workers and that that inequality has increased since 1970.

Some policies narrow inequality and some widen it. Again and again we will see that the basic dimensions of social inequality—how rich the rich are and how poor the poor are, and even who becomes rich or poor—are a result of our social and political choices. Many of our policies operate in-directly, and hence invisibly. The programs that help the poor are glaringly obvious, but those that aid the rich and middle class tend to be invisible. Obscured even more are the policies that set the rules of "the game" for the labor market. In this chapter, we will reveal some of the many ways that social policy shapes inequality.

We will begin by looking at one general pattern of American social pol-icy, which is to provide, with one hand, limited direct help to some of the poor and indirectly to subsidize, with another, the middle class and the wealthy. Next, we will uncover one of the most hidden arenas of social policy, the regulation of the labor market, and show how the ground rules shape inequality. Finally, through an examination of higher education, we

BASEBALL: HOW RULES HELP PICK THE WINNERS

In sports, talent and effort should determine who wins. But the rules also determine who has the advantage and even the kinds of players who succeed. We see that in the history of baseball. In 1893 the team owners lengthened the pitching distance to sixty feet and pitchers lost an edge; runs soared to an average of fifteen per game the next year. Then, in the early 1900s, the number of runs per game plummeted (to seven in the National League) because new rules created a wider home plate and counted the first two foul balls as strikes. These new regulations, in turn, partly determined who became successful players. Baseball historian Benjamin Rader writes:

> With the lengthening of the pitching distance in 1893 and the growing practice of placing the pitching slab on a mound of dirt ("the mound"), the sheer size of pitchers began to increase sharply. In 1894 the pitchers were relatively small; they averaged 168 pounds (4 pounds lighter than the hitters) and stood at five feet and ten inches tall, the same height as the hitters. By 1908, however, pitchers (at 5′11″) averaged an inch and a half taller than the hitters and were an average of 9 pounds heavier (180 lbs. vs. 171 lbs.). Notice that the pitchers of 1908 weighed a whopping 12 pounds more than the average of their counterparts in 1894.

> Following another hitting drought in the mid-1960s, baseball owners again attempted to right the balance by lowering the pitching mound from fifteen to ten inches, ordering umpires to tighten the strike zone, moving fences closer to home plate, and in the American League allowing a "designated hitter" to bat for the pitcher.* As we write, owners once again are changing rules in the pitcher's favor.

> * Rader, *Baseball*, pp. 87, 89, 114–16, 169.

will look at some of the diverse ways in which public investment also molds inequality. In the end, we will better understand the major reasons why inequality is historically so inconstant and why inequality in America is so high.

VISIBLE POLICY: REDUCING POVERTY
THROUGH REDISTRIBUTION

Over the last century, American government has done much to help those left poor by the market. Public health programs, school lunches, food stamps, Aid to Families with Dependent Children (AFDC), and survivors' benefits have reduced the inequality left by earnings differences. Yet Americans have chosen not to pursue such programs as far as citizens in other affluent nations have (and the programs are being sharply cut back as we write). Most industrial societies provide "family allowances" to all families with children and some form of universal health care or health insurance to all residents. In such ways, the numbers and problems of the very poor are sharply reduced by government policies that are directed toward everyone and that do not single out the poor. Most American welfare programs, in contrast, are "means-tested"—available only to those who can prove that they are poor and that they are otherwise deserving. These targeted programs consequently lack wide political support and are vulnerable to budget-cutting. Only social security and Medicare, nearly universal entitlements for the elderly, have largely survived cutbacks in recent years. Most other nations, unlike the United States, also substantially subsidize housing for many moderate-income citizens, provide stipends for students who make it into higher education, and support the long-term unemployed.

Recent American antipoverty programs have had some success, but mostly in reducing poverty among the elderly, largely through social security and Medicare, and in taking the edge off misery (see figure 5.5). We can see the emphasis on the elderly by looking at the percentage of Americans who are pulled above the poverty line by all government financial programs (taxation, unemployment support, welfare, social security, etc.) put together. In 1992, 22 percent of Americans would have had incomes below the poverty line if all that had been available to them were their families' earnings. Government taxes and transfers reduced that to 12 percent, a drop of ten points in the proportion of poor Americans. For the elderly, taxes and transfers reduced the proportion by *forty points*, from the 50 percent who would have been poor based on nongovernmental income alone to the 10 percent who were poor after including governmental income and taxes. For children, however, the net effect of taxes and transfers was to reduce poverty rates by only *seven points*, from 24 percent to 17 percent. For young adults, the drop was merely five points, from 21 percent

131

to 16 percent.[1] This generational imbalance is, in part, the outcome of policy changes during the 1980s that weakened the equalizing effects of taxes and transfers.[2] (As of yet, we have no data on the effects of the 1993 Clinton tax changes that raised the earned income tax credit for low-income families and raised the income tax rates for the very wealthiest households. Presumably, these laws shifted net incomes toward equality a little. But the changes enacted by the Republican Congress elected in 1994 will shift incomes away from equality.)

If we list all the programs that helped nonelderly Americans with low income—food stamps, AFDC, Women, Infants and Children (WIC—a nutrition program), Medicaid, SSI disability, the earned income tax credit, etc.—they sound like a lot. Adding together these programs and adding in as well a variety of federal, state, and local spending directed not just at the poor but also at many people who are above the poverty line, such as college loans, job retraining, and energy assistance, the total expenditures for "persons with limited income" in 1992 amounted to almost $290 billion. As sizable as that figure is, it represents less than 12 percent of all government expenditures at all levels that year. It comes to about $5,900 per low-income person. Almost half of this total, $134 billion, was for medical care, largely Medicaid. Nonmedical spending came to about $3,200 per limited-income person, of which about $2,100 was in the form of cash or food stamps. That $2,100 is roughly what the typical American family spent on eating out in 1992; it is within a few hundred dollars of what typical home-owners saved on their federal income taxes by being able to deduct mortgage interest. Even after this government spending—which is probably a high-end estimate of what America spent to aid low-income people in 1992—over 14 percent of Americans, 21 percent of American children, remained poor.[3]

We can best evaluate the effort to redress poverty comparatively. Low-income American children are worse off than low-income children in any other industrial nation. In the 1980s, for example, about 20 percent of American children lived in poverty, while 9 percent of Canadian children and of Australian children were poor, 7 percent of children in the United Kingdom, and even fewer in France, West Germany, and Sweden, respectively.[4]

Why are so many American children poor? Charles Murray claimed in an earlier book that American children are poor because welfare policies encourage poor women to have more children. He is wrong. Careful studies by demographers demonstrate minimal effects, if any, of AFDC on child-bearing. Rather, young parents are more susceptible to poverty, and their poverty makes their children poor.[5] American children are more often poor, first, because American adults are more unequal in both wealth and income than people in any other industrial society. Second, children suffer especially because the incomes of young men have fallen so sharply since the mid-1970s. More young men cannot earn enough to keep their children out of poverty, and many then refuse to take on the responsibilities of marriage, leaving young mothers and children even poorer.

We can see how American government compares with others in dealing with poverty by turning again to the Luxembourg Income Study. Lee Rainwater and Timothy Smeeding calculated, for eighteen nations, the percentage of children who were poor. (To be able to compare across countries, "poor" was defined as being in a household with real purchasing power less than half that of the median in the nation. Half the median is roughly what the poverty line in the United States was in the 1960s when it was first calculated.)[6] Figure 6.1 shows the percentage of children who were poor before and then after including taxes and government transfer payments in the calculations. Again, we look only at the populous nations. Before government intervention, a relatively high percentage of American children were poor, but not as high as in France and the United Kingdom. After counting taxes and government payments, however, the poverty rate for American children was substantially higher than that elsewhere (including nine other nations not shown in the figure). Even those countries with higher before-government child poverty than the United States managed to reduce their poverty levels to far below the level here.

Two objections might be raised to the evidence that America leaves so many of its children in poverty. One is that so many American children are poor because so many live in single-parent families. That is true. However, Rainwater and Smeeding also looked separately at children in two-parent and in single-parent families. In each case, the same pattern appears as in figure 6.1: American children were exceedingly likely to be left poor after government action.[7] The other objection is that being poor in America, being below 50 percent of the median, is in material terms not as terrible as being poor elsewhere. Unfortunately, that is not so. Rainwater and Smeeding calculated how much real purchasing power children at the 10th, 50th, and 90th percentiles of the income distribution had available in each country. American children near the top and at the middle did, indeed, have more real income than did children near the top and at the middle elsewhere. But American children near the bottom had *less* real income than children in the other nations, 25 percent less than poor Canadian children and 40 percent less than poor West German children. And again, the

Government in the United States does the least to reduce the proportion of children who are poor: percent of children poor before and after government action.

6.1. Percentage of Children Who Are Poor, Before and After Government Action, in Eight Nations (*Note*: "Poor" is defined as children in households with incomes below 50 percent of the national median household income. Government action includes all taxes and all cash and "near-cash" transfers. *Source*: Adapted from Rainwater and Smeeding, "Doing Poorly," table A-2)

researchers did not count some of the in-kind resources provided to poor children overseas.[8]

The United States does less than any other advanced nation to reduce poverty through government benefits.[9] In addition, our spending to aid the poor is precarious. Assistance is a donation; it is not a right, as it is elsewhere, and it is therefore politically vulnerable. For example, the main program that supports children, AFDC, has been repeatedly cut such that the monthly benefits dropped 25 percent in real value between 1980 and 1993.[10] And it will continue to decline as the federal government transfers responsibility for welfare to the state governments. Also, unlike aid to the poor in most other nations, these programs form an overlapping, conflicting, and sometimes impenetrable morass. One reason they do is that each program was targeted to a specific need and requires would-be clients to meet exacting qualifications. These features of American policies to help the unfortunate are consistent with Americans' belief that aid to the poor must be limited, conditional, and sufficiently unappealing so as to push people off assistance and into jobs.[11]

One of the newest and most ambitious efforts to redistribute income, one consistent with the American attitude toward aid, was the expansion of the earned income tax credit (EITC) that President Clinton, with bipartisan support, enacted in 1993. The EITC was originally established to refund low-income earners the money deducted from their paychecks to pay for federal insurance programs like social security. The 1993 expansion turned it into a general income support program for such workers, in which earners would get more money from the federal government the more they earned on their own, up to a ceiling. According to the 1993 law, workers filing income tax returns would have gotten, by 1996, credits worth 40 cents for every dollar they earned, up to a maximum refund of about $3,500. (That would have been for a family of four with earnings in the $8,400–$11,000 range; a fully employed worker earning the minimum wage grosses under $9,000.) Beyond $11,000 in earnings, the EITC would gradually shrink until reaching zero for families earning $27,000. Expansion of the EITC was initially popular among conservatives because it rewards working and among liberals because it provides the working poor with supplementary income in a nonstigmatizing way (applicants need only fill out an income tax return). Projections were that by the year 2000, EITC would have transferred $30 billion a year to poor and low-income families, more than either AFDC or food stamps.[12] However, the changes enacted by the 1995–96 Congress scaled back the expansion of the EITC.

The history of the EITC sheds light on the American approach to redistribution. First, the EITC provides no support at all for families whose head of household is, for whatever reason, not working. One must earn and *deserve* the help; the more you get in the marketplace, the more you get from government. Second, more generous than most programs, it is still limited. By one estimate, had it been fully implemented in 1996, EITC would have moved only about 25 percent of *working* poor families above the poverty line. Third, the design and discussion of the EITC has largely focused, not on how best to reduce poverty, but on how to aid only the deserving and to reward work effort. That is why, fourth, it encountered serious political trouble. Anger at unqualified recipients who fraudulently claimed the credit and concerns that the rebate structure may lead some workers near the top of the eligibility range to cut back on their work hours—together with efforts to balance the federal budget—propelled Congress to scale back the EITC.[13]

Overall, then, American government policy *does* reduce inequality by aiding the poor. The New Deal and the Great Society programs substantially helped the elderly and reduced some of the misery for others—all that the programs were ever designed to do.[14] This mix of policies may be what most American voters wish. There are prominent voices arguing that even this amount of redistribution to the nonelderly poor is too much, that it hurts the economy and even hurts the poor by undermining their self-reliance. Our point here is that in the case of our most visible policies, ones to aid the poor, Americans have *chosen* to do less rather than more, have designed greater inequality.

INVISIBLE POLICIES I:
SUBSIDIZING THE MIDDLE CLASS

In contrast to the highly visible, if limited, direct aid given to the poor, American social policy tends to subsidize the middle class more generously, but indirectly and less visibly. The effects of these indirect policies have generally been to simultaneously *decrease* inequality between the middle class and the wealthy and to *increase* the gap between the middle class and low-income Americans.

Subsidizing Homeownership

The mortgage interest deduction is a quintessential example of the invisible ways American policy subsidizes the middle class and the wealthy and, indeed, offers a greater benefit the wealthier one is. A person too poor to buy a house receives no housing subsidy (unless he or she is so poor as to qualify for welfare), while a wealthy homeowner with a mansion and a vacation house may receive a subsidy worth tens of thousands of dollars. For example, someone carrying a million-dollar mortgage would get tax breaks worth over $33,000 *a year*.[15]

It might seem odd to think of the mortgage interest deduction as a "subsidy," because it is the taxpayer's own income that he or she keeps. But a subsidy it is, because it is a tax that the government forgoes in its effort to encourage homeownership. Had the taxpayer with the million-dollar mortgage rented the same home instead of buying it, he or she would have had to pay $33,000 more a year in taxes. If United States tax policy treated mortgage expenses the same way it treats other living expenses, like food, rent, cars, or clothes (for which no deductions can be taken), the government would have far greater revenues. Policy experts thus refer to deductions like mortgage interest as "tax expenditures," an awkward term, but one that accurately indicates that tax deductions cost the government money—which is to say that *they cost other taxpayers money*. Whatever one person saves on taxes, others must make up in taxes, or government debt, or reduced government services. By the early 1990s, the cost of the mortgage interest and property tax deductions amounted to more than $60 billion annually, over four times as much as was spent on direct housing assistance for low-income families.[16] (This expenditure was untouched in budget-balancing legislation of 1995–96.)

And even these figures do not fully measure the subsidy homeowners receive from the government. The government underwrites much of the real estate industry by insuring and regulating private mortgage lenders. Before the Great Depression, when the government first began guaranteeing and monitoring home loans, banks typically required a 50 percent down payment on homes and normally issued mortgages for only five or ten years. Not surprisingly, under those conditions, homeownership was beyond the means of many middle-class families. Only with government intervention in the housing finance system, through the Federal Housing Authority (FHA) and the Veteran's Administration (VA), did thirty-year mortgages at relatively low interest rates become common. And these long-term, low-rate mortgages, along with the mortgage tax deduction, are what has enabled so many middle-class Americans to buy their own homes.[17]

Another government aid to homeownership has been its massive road construction effort. For many years, but most especially in the 1950s, the federal government began ambitious projects to build tens of thousands of miles of new highways. Partly because of these highways, 85 percent of the new housing built after World War II was erected in the suburbs, where land was plentiful and relatively inexpensive.[18] Many of the new highways connected the suburbs to the downtown business districts of large cities, making it possible for middle-class Americans to buy a home in the suburbs and commute to work. Yet other government policies contributed to

How Am I Subsidized? Let Me Count the Ways

Americans pride themselves on their independence. We admire the "self-made" millionaire, and we all like to think that we achieved our success on our own. These may be admirable values, but they can lead us to misunderstand our own lives and to be harsh toward the less fortunate. What would happen if a middle-class American, someone like the authors or the likely readers of this book, simply tried to count all the ways he or she was "subsidized" by the larger society?

All of us born in the United States were indirectly subsidized at birth. The public health measures that reduced disease, that eliminated smallpox and nearly wiped out polio, that provide us with safe drinking water, and that regulate hospitals and medical practice have been provided by public agencies. Then, in childhood, all of us received a vast subsidy in the form of public schooling.

Even those things we "earn" through our own effort or ability are often subsidized directly and indirectly. College education, for example, is subsidized for most people. Almost all public universities and most junior colleges charge much less in tuition than a student's education really costs. And most top private universities spend money generated by their endowments (accumulated from tax-deductible gifts and bequests) to provide students with a more expensive education than even full tuition would pay for. Only proprietary schools, like barber colleges and secretarial schools, and private colleges without endowments charge students as much as it actually costs to educate them (and even some of them receive indirect subsidies).

Businesspeople who employ highly educated workers, such as graduates of state engineering schools, receive a subsidy because their employees are trained at public expense. Businesses also depend on the indirect subsidy that supports the physical infrastructure, from roads and bridges to sidewalks and parking spaces.

These subsidies, as well as the mortgage deduction and the tax subsidy for health insurance, are for the most part to the good. Societies are joint endeavors in which each of us necessarily depends on many others, sometimes directly and often indirectly through the provision of public services ranging from national parks to sewer lines. One difficulty, however, is when these subsidies largely help those already well-off, such as the mortgage subsidy. Another difficulty comes when we imagine that we stand alone; that no one has ever done anything for us; and that we owe nothing to anyone else.

the expansion of middle-class homeownership in the postwar decades. Federal GI benefits allowed veterans to purchase their homes with a single dollar down. And government-funded research provided the plywood paneling, aluminum siding, and prefabricated walls and ceilings featured in the affordable housing of the 1950s and 1960s.[19]

Before World War II, many fewer Americans owned homes than do today. Nonfarm homeownership was confined primarily to the affluent. It was government policy that brought homeownership to large numbers of middle-class Americans.[20] Homeownership gave these middle-class Americans independence, real property to pass on to their children, and an opportunity to make the kind of financial investment that once only the wealthy had been able to afford.

Government subsidy of homeownership is a much-applauded policy that has shrunk some of the gap in the standard of living between wealthy Americans and middle-class Americans. But it has not yet worked that way for lower-income Americans who cannot afford to enter the housing market at all. Also, in recent years, this system has worked more to the advantage of the especially wealthy. After World War II and into the 1970s, tax expenditures for mortgage interest deductions helped the middle class about as much as they aided the wealthy. By the late 1980s, however, these tax expenditures benefited those at the top of the income distribution far more than those in the middle. The most recent statistics show that 44 percent of the mortgage subsidy goes to the 5 percent of taxpayers with incomes above $100,000 a year and that half of all homeowners receive no deductions at all.[21]

Other affluent nations have gone further than the United States in equalizing housing. Like the United States, they subsidize the middle class and the wealthy through tax deductions for mortgage interest payments. But elsewhere tax deductions are a part of comprehensive programs that also include relatively generous housing support for low- and moderate-income households.[22] In most European countries, for example, governments provide rent assistance for many working-class and poor families,[23] and in many of these countries governments also finance the construction of new housing.[24] By providing subsidies to a larger percentage of the population, and especially to those at the lower end of the income scale, European governments have tended to lessen inequality across the board.

Housing and Discrimination

American policies that promoted middle-class homeownership decreased the distance between the wealthy and the middle class. But these government programs, which made such an enormous difference to the security and well-being of generations of Americans, were essentially denied to black Americans.

Before the federal government would guarantee a loan through the FHA or the VA, it required a professional appraisal. Appraisers always rated black neighborhoods in the lowest of the four possible categories (indicated by the color red on the maps used by federal appraisers—hence the term "redlining") and usually rated neighborhoods near a black district in the next-to-worst category. Either designation was enough to render a property ineligible for FHA- or VA-guaranteed loans.[25] Other policies directly blocked African Americans from moving into white neighborhoods. Berkeley sociologist Troy Duster notes that "in 1939, the Federal Housing Authority's manual . . . stated that loans should not be given to any family that might 'disrupt the racial integrity' of a neighborhood. Indeed . . . the FHA manual went so far as to say that 'if a neighborhood is to retain stability, it is necessary that properties shall be continued to be occupied by the same social and racial classes.'" By the late 1940s, the FHA was recommending that developers use racially restrictive covenants as a way to ensure the financial viability of neighborhoods.[26]

Duster goes on to note that as a result of these policies, whites were able to get government-supported mortgages at 3 to 5 percent interest, "while blacks were routinely denied such loans. For example, of 350,000 new homes built in Northern California between 1940 and 1960 with FHA support, fewer than 100 went to blacks." By 1962 the VA and FHA had financed more than $120 billion in new housing, but less than 2 percent was available to nonwhite families, and most of that only in segregated neighborhoods.[27]

Other housing and development policies also widened class and racial divisions. Loan policies favored the expansion of suburbia, as did the building of the interstate highway system. The suburbs welcomed the jobs and stores that moved out with middle-class whites, while setting financial and racial barriers to city blacks who might try to move out. As a result, most blacks remained in urban ghettoes, far from growing centers of employment. Also, federal and local governments placed large-scale public housing projects, which concentrated the poor, in those same redlined neighborhoods. The contemporary isolation, concentration, and separation of poor blacks in inner cities is not simply the result of market forces, or even of market forces in combination with private racial discrimination. Rather, government policies directly contributed to widening inequality between those who were able to buy homes and move to the suburbs during the postwar housing boom (almost exclusively whites) and those who were too poor to receive government help or who, like most minorities, were excluded from programs that subsidized new housing.

The ramifications of discriminatory policies and practices go beyond housing and segregation. Housing is most Americans' major financial asset; it can be used to leverage credit from lenders; and it can be the major inheritance left to the next generation. While almost two-thirds of white households have home equity, at a median value of $45,000, only two-fifths of black households do, at a median value of $31,000. Blacks who have graduated from college earn 76 percent as much as whites who have graduated from college, but they have only 23 percent as much net worth.[28] Part of the reason for these gaps is the legacy—and continuation, too (see chapter 8)—of housing discrimination.

Health and Health Care

Health care is another arena in which social policy invisibly benefits better-off citizens while neglecting the needs of many of the less advantaged. Critics have noted how much Americans now spend on health care—more per person than any other nation.[29] Less often noticed is that the dominant way we provide health care here—private insurance, most often through plans offered by employers—tends disproportionately to subsidize high earners. As one expert notes, the tax system encourages health insurance through employers because employers' share of the insurance premiums does not count as part of employees' taxable income. But those who are too poor to pay any income taxes at all receive no subsidy for health insurance. They are therefore unlikely to be insured, because each dollar diverted to health insurance represents a full dollar less that they, lacking any tax break, might have gotten in wages. The estimated 1992 value of the federal tax subsidy—the untaxed employer contribution—was $270 for households in the bottom 20 percent of the income distribution, $525 for those in the next 20 percent, and $1,560 for those in the top 20 percent.[30] No wonder most of the poorly paid go without insurance even when they are employed—half of fully employed poor Americans lacked health insurance in 1993[31]—while most of those in high-paying jobs are insured through their employers.

Health policy has improved the lives of some of the poor. Since the 1960s, elderly Americans and the disabled who receive social security benefits have been insured through Medicare, the federal health care program, while Medicaid, a joint federal-state program, pays for the health care of

PARADOXES OF WAR: MORE EQUALITY IMPROVES HEALTH

The distinguished Harvard economist Amartya Sen has demonstrated the remarkable effects of social policy on health, even in times of general shortage. He shows that life expectancy in England increased throughout the twentieth century, but the two big advances in length of life occurred during the decades that included World War I and World War II. The reason was "dramatic increases in many forms of public support including public employment, food rationing and health care provisions."* That is, during wartime shortages food and health services were more evenly shared.

Longevity Expansion in England and Wales:
Increase in Life Expectancy per Decade (Years)

	Male	*Female*
1901–1911	4.1	4.0
1911–1921	*6.6*	*6.5*
1921–1931	2.3	2.4
1931–1940	1.2	1.5
1940–1951	*6.5*	*7.0*
1951–1960	2.4	3.2

Despite the deprivations of war, including food shortages and food rationing, there were dramatic improvements in public health when the English and Welsh had more equal access to what is available. Generally, a nation's life expectancy improves as it becomes more economically equal**—another instance in which policy affects inequality in its most profound sense.

* Dreze and Sen, *Hunger and Public Action*, pp. 181–82.
** Williams and Collins, "US Socioeconomic and Racial Differences in Health."

some disabled and low-income persons. As we noted earlier, Medicaid accounts for roughly half of government spending on nonelderly low-income people. In 1990 Medicaid covered about 18 percent of all American children and about 61 percent of poor children.[32] Barbara Wolfe notes that before Medicaid, "children in families with high incomes saw physicians 67 percent more often than children in low-income families," while after the introduction of the Medicaid program, "children in families with high incomes saw physicians only 20 percent more often than children in low-income families."[33] These are widely supported programs that have measurably reduced substantive inequality.

However, this medical coverage is so spotty that it leaves many uninsured. Young adults (who are not covered by Medicare and are unlikely to be covered by Medicaid), low-income earners not quite poor enough to qualify for Medicaid, and especially blacks and Hispanics are likely to be among America's 37 million uninsured. Among low-income families above the poverty line, more than one in five remains uninsured all year.[34] This very uneven distribution of access to health care has real consequences. Adults in the poorest fifth of the population have more than three times the number of health conditions that limit their activities (including work) as those in the top half of the income distribution.[35]

The United States and South Africa are the only two countries in the industrialized world that fail to provide medical insurance for all their citizens. Universal coverage, of course, reduces inequality in standards of living. The system in place in the United States is more capricious and less equalizing. Americans who work for employers offering medical insurance, or who are over sixty-five, or who are so poor that they receive welfare all have subsidized access to health care. But those Americans who work for employers who do not provide health insurance, most of whom are younger, poorer, and darker skinned, receive inadequate health care and often are in worse health as a result.

Subsidizing Families

Family policy is another realm in which social policy quietly supports middle-class and wealthy Americans while providing limited benefits to poor Americans. The primary way the United States has supported families is through *tax deductions* for children. Through 1995 the deduction amounted to about $2,400 for each dependent child, which was worth about $750 in a tax refund per child for affluent families who were in the 31 percent income tax bracket. It was worth less to families in lower brackets, and the deductions were worth nothing to families that are too poor to pay income taxes.[36] (The child tax credit passed by the 1995–96 Congress is less skewed toward the wealthy because it is worth the same to all taxpayers. But it still amounts to nothing for families too poor to pay taxes.) The earned income tax credit discussed earlier redresses this imbalance only partly.[37]

In other advanced countries, flat cash allowances are paid instead. Rich and poor families receive the same amount per child. Of course, those amounts make a greater difference in the lives of poorer families.

In the United States, even child-care support flows disproportionately to the affluent, in the form of the dependent care tax credit. This tax credit, as of 1995, allows families to use pretax money for child care, up to a tax savings of $1,440. As with other tax-based federal subsidies, those who earn so little that they pay little or no income taxes receive no benefit. In 1994 "Depcare" cost about $2.8 billion in forgone taxes, close to the $3.3 billion budgeted for Head Start, the school-preparation program targeted to poor children.[38]

As with homeownership and health care, American policy for children narrows economic inequality in America, especially between the middle class and the wealthy. This confirms that inequality is within our control. However, with children, health care, and homeownership, American policy remains considerably less equalizing than policies elsewhere in the developed world. Many American policies, in fact, subsidize the affluent instead of those with lower incomes. Most of these subsidies for the middle class and higher are nearly invisible, while those to the poor bear the flashing neon light, "WELFARE."

INVISIBLE POLICIES II: SUBSIDIZING THE WEALTHY

Another set of subsidies also often goes unremarked: those that directly or indirectly help the very wealthy. While policies that help the middle class reduce inequality between them and the rich, these subsidies clearly increase inequality altogether by elevating the rich above everyone else.

Corporate Welfare

American public policy promotes inequality through tax breaks and subsidies for corporations. While newspapers report sensational details like $2.9 million to help the Pillsbury Corporation advertise abroad and $263,000 for a Smokey Robinson concert given by Martin Marietta at taxpayers' expense,[39] direct subsidies and tax breaks to corporations take such myriad forms that it is hard to measure their full extent. Perhaps the best-known subsidies are those to farmers. These are programs that support the prices of basic commodities, pay farmers not to plant some of their acreage, or subsidize the price of farm products exported abroad. While the programs were designed originally to help low- and moderate-income family farmers, large-scale farmers and agribusiness are the big winners. The Progressive Policy Institute, in a report joined by the libertarian Cato Institute and the conservative Heritage Foundation, estimated that cutting, not eliminating, such agricultural subsidies would save $31 billion over five years.[40] A second major set of subsidies goes to energy producers and other natural resource firms. Federally owned hydroelectric plants sell electricity to utilities at below-market rates; the Forest Service builds roads into national forests for the timber industry; and the federal government finances research for the nuclear and fossil fuels industries that these industries could pay for themselves.[41]

Overall, the Cato Institute estimates that "at least 125 separate programs providing subsidies to particular industries and firms" cost taxpayers some $85 billion per year.[42] Analysts recognize the need for vital national investments—building highways and bridges, constructing irrigation systems that aid agribusiness, supporting research and development in start-up industries, funding mass transit. But they argue that most existing subsidies soak up resources that would otherwise go to productive investment, public or private. Current federal subsidies to industry are mostly historical legacies of no-longer-pressing problems (like the subsidies to miners and cattle ranchers meant to encourage settling the West) or are responses to lobbying by powerful interest groups.

The federal government *indirectly* subsidizes many industries by providing free regulatory services that are crucial to doing business. The Federal Aviation Administration provides such vital functions as the system of air traffic control to airlines as a free good. It is a subsidy to their business operations (and indirectly to airplane travelers). Coast Guard rescue and enforcement activities subsidize commercial boat companies and pleasure boat operators. The Securities and Exchange Commission's fees charged to financial firms do not cover the full costs of registering and monitoring securities transactions. Businesses using inland waterways do not pay what it costs the federal government to maintain and operate them.[43]

Industry subsidies also operate indirectly through special exemptions and deductions written into tax codes. The best known is the oil depletion allowance, which allows oil and natural gas companies to deduct a percentage of their gross income for tax purposes. (Unlike other industries, oil, gas, and mineral firms are also allowed to deduct fully some of their capital costs, rather than depreciating them over the life of the investment.) It is estimated that reforms to eliminate these and many other tax breaks for oil, gas, and mineral producers, the financial industry, the construction industry, agribusiness, the timber industry, and many others would save another $101.8 billion over five years.[44]

In even less obvious ways, the federal government supports the infrastructure that maintains the livelihoods of the wealthy—from maintaining the regulatory apparatus that keeps the stock market functioning smoothly to the enormous costs of the savings and loan bailout. One might argue that these expenditures are good for the economy as a whole and ultimately good for everyone. That may well be so, but they do redistribute wealth upward. In the end, they most benefit the shareholders of the corporations that are subsidized. Since the highest-earning 5 percent of families own over half of American corporate stock but pay less than a third of all the taxes, they are the big winners in this redistribution—by tens of billions of dollars.[45]

Taxes

Recent changes in tax laws are another set of political choices that have helped push inequality in America to a level that leads the West. The great investigator of recent tax reforms and their effects on inequality is not a left-wing radical but Republican commentator Kevin Phillips, concerned that resentment of growing inequality will produce a populist backlash. In a pair of deliberately unsettling books, *The Politics of Rich and Poor*, published in 1990, and *Boiling Point*, published in 1993, Phillips drew a devastating portrait of changes in the fate of the American middle class between the postwar period and the present.

Phillips traces the explosion in income inequality to the decade of the 1980s. Between 1977 and 1990 the average income, in constant dollars, of the bottom tenth of American families fell by 11 percent, while the average income of the top tenth rose 20 percent to $133,200. (See table 6.1; it presents in greater detail the general trend noted in figure 5.2.) Even more dramatically, the income of the top 1 percent of families rose 45 percent to $463,800 by 1990.[46] Most American families fell behind in real income—despite the fact that more and more wives began bringing home a paycheck—while the very wealthy made spectacular gains.

How did this happen? Phillips points to the "soak the middle" effects of the early Reagan administration's tax changes. While the first Reagan-era tax cut of 1981 reduced the average family's federal income taxes by 25 percent over three years and indexed them against inflation, increases in the social security tax and then the 1986 "tax simplification" package in-

TABLE 6.1

Between 1977 and 1990 the Family Income of the Bottom 60%
of American Families Declined, While That of the
Top Groups Increased

| | Average Income Level (1988 dollars) | | Percent Change |
Decile	1977	1990	1977–1990
First	$4,277	$3,805	−11.0
Second	8,663	8,251	−4.8
Third	13,510	13,110	−3.7
Fourth	18,980	18,200	−4.1
Fifth	24,520	23,580	−3.8
Sixth	30,430	29,490	−3.1
Seventh	36,880	36,890	0.0
Eighth	44,820	46,280	3.3
Ninth	56,360	59,860	6.2
Tenth	111,100	133,200	19.9
Top 5%	149,500	187,400	25.4
Top 1%	319,100	436,800	45.4

Source: Phillips, *Boiling Point*, p. 28. Data from Congressional Budget
Office, House Ways and Means Committee, 1992 Green Book.

Note: Income includes capital gains income.

creased the tax burden on the middle class, while substantially decreasing
taxes on the wealthy.[47] Despite what were hailed as tax cuts, the *effective*
federal tax rate (income tax plus FICA) for the median American family
increased from 23.7 percent in 1980 to 24.6 percent in 1990. Meanwhile,
the effective federal tax rate for the highest-earning 1 percent fell dramatically, from 35.5 percent in 1980 to 26.7 percent in 1989. The 1986 Tax
Reform Act saved the family in the $10,000 to $20,000 tax bracket $69 in
taxes and saved the family in the million-dollar-plus bracket $281,000.[48] A
longer historical view shows an even more dramatic shift: The steeply progressive federal tax system that was in place from World War II until
1970—millionaires in 1960 had an official federal tax rate of more than 85
percent, although many loopholes made that only a nominal rate—became
nearly flat in the late 1980s.[49]

Federal tax and spending cuts also had two indirect, unequalizing results. First, cuts in federal spending shifted burdens to state and local governments, whose taxes and user fees increased sharply. The percentage of

average families' incomes going for state and local taxes increased from 9 to 10 percent, and the relative burden on poorer families increased even more.[50] Subsequent cuts in local government meant sharp cutbacks in such amenities as library hours, road repair, police patrols, and primary education; steep increases in user fees for everything from junior college to bus fares and garbage pickup put further stress on the pocketbooks of the middle class and the poor.

Second, the most striking impact of changes in federal tax policy was to increase the *pretax* incomes of the very wealthy. In the early 1980s, the maximum tax rates on income from investments (capital gains and unearned income) dropped sharply. This increased the net value of stocks, bonds, and other financial assets, sparking a boom in the prices of such assets and the income their owners received. This in large measure explains why the richest 1 percent saw a 90 percent growth in their incomes in the 1980s. Between 1978 and 1988 the number of individuals with million-dollar annual incomes soared from 2,041 to 65,303.[51] Millionaires did not multiply overnight because there was a sudden increase in individual talent. Political choices in the 1980s, in large part choices about tax policy, reshaped inequality in America.

(At this writing, the tax law changes likely to go into effect in 1996 point toward increased subsidies of the wealthy.)

Invisible Policies III:
Regulating the Labor Market

Thus far, we have seen that how the government collects and spends tax dollars affects inequality in America. But social policy begins shaping inequality long before the tax bills come due and the social security checks are mailed out. The market itself is structured by policy choices, by how we set the "rules of the game." We noted this briefly in chapter 5, pointing to rules such as licensing requirements and laws protecting corporations. Here, we will take a closer look at how ground rules help determine who wins and who loses and how much they win and lose.

Consider, again, the large increase in income inequality that occurred in America in the 1980s. The real losers were less-educated workers. They lost ground not only relative to their college-educated peers, as we have seen, but also compared with their counterparts a decade earlier (see chapter 5). Market forces alone cannot explain the increase in inequality in the 1980s. If so, one would expect other advanced industrial countries to have experienced similar increases in inequality in the 1980s. They participate in the same world markets, use similar technologies, and have similar types of industries and occupations. Yet these other countries experienced neither the same large increases in wage inequality nor the drops in the real earnings of the less skilled. In Canada, Japan, and Sweden wage inequality grew, but much more modestly than it did in the United States. In France and Italy, inequality changed hardly at all.[52] (Only in Great Britain did the gap between professionals and blue-collar workers increase as much as it did in the United States, and there only because of gains at the top; the pay for people with low wages rose in Great Britain, just not as quickly.) Low-wage workers in the United States now earn only about half as much relative to American high-wage workers as low-wage workers in Europe earn relative to high-wage ones there.

Why did other advanced industrial countries experience less wage inequality in the 1980s than the United States did? Because, in large part, they made different policy choices. In particular, other advanced countries have different rules about unionization and have different wage-setting institutions.[53]

Union Rules

Economists Richard Freeman of Harvard and David Card of Princeton estimate that the sharp decline in the percentage of unionized workers in the 1980s explains at least one-fifth of the growth in wage differentials among male workers in the United States.[54] This is because unions reduce the pay gap between higher- and lower-ranking workers.[55] In the United States in the 1980s, there were simply too few union members to offset growing inequality. Between 1970 and 1990 the proportion of the labor force that was unionized dropped more than 45 percent to only 11 percent of the private sector, virtually the lowest unionization rate in the industrialized world.

Unionization, in turn, has declined so precipitously in the United States largely because of the unusually hostile political and legal environment here. Especially instructive is the comparison with Canada, because Canada and the United States share similar cultures, economic institutions, and standards of living. In the 1940s Canadians revised their labor laws to resemble the United States' 1935 Wagner Act, which established legal procedures for labor organization and collective bargaining here. Since then, however, Canadian labor laws have become more favorable to unions while American labor laws have become less favorable.[56] Under current

Canadian law, a union is established once a majority of workers sign a card indicating their support. Under current American law, after a majority of workers have signed such cards, unions must still go through a long election campaign, often facing management consultants hired by employers to convince workers that they do not want a union after all.[57] Also, it is illegal in Canada for employers to replace strikers permanently, but it is permissible to do so in the United States. This was brought to the consciousness of many Americans during the baseball strike that ran from August 1994, to March 1995. The Canadian government forbade foreign replacement workers, and Ontario provincial law prohibited hiring any replacements at all. The Toronto Blue Jays were forced to schedule their possible 1995 "replacement baseball" season in Florida.

In the 1950s and 1960s, when labor laws and practices were most similar in the two countries, unionization rates were also similar, but since then unionization rates have risen in Canada and dropped sharply in the United States. By the early 1990s the percentage unionized in the private sector in Canada was almost three times larger than in the United States.[58] Partly because of these differences in unionization rates, wage inequality grew much less in Canada than in the United States in the 1980s.[59]

Unions and Plant Relocation

In the 1980s some major American companies busted their unions in celebrated cases (after President Reagan had defeated the air traffic controllers' union). More often, however, employers escaped union pressure by moving from one state to another or out of the United States altogether. The scale of movement during the decade of the 1980s alone is staggering. University of North Carolina sociologist John Kasarda estimated that the northern and midwestern states lost 1.5 million manufacturing jobs and $40 billion in pay between 1980 and 1990.[60] One-third of the jobs ended up in southern and western states; some of the rest moved overseas; some were lost to automation; and some were simply lost as firms stopped producing goods.

The competition among localities for jobs-on-the-move is intense. (The struggle to land sports franchises is a vivid illustration.) The competition among cities and states for firms usually turns on tax concessions, capital commitments, and promises to regulate union activity.[61] But the costs to the victors are significant. They do not necessarily increase their tax bases, because the bidding frequently requires giving away tax revenues; also, local taxpayers often contribute to firms' relocation costs. National policy allows states to differ greatly in laws protecting labor and thus to compete on the basis of who has the weakest ones, thus encouraging the shift of jobs to weakly unionized—and lower-paying—states.

Wage Setting

In the United States, workers' wages are negotiated either by an individual employee with his or her employer or by a local union with a specific employer. In many jobs, the employer simply offers the job at a preset wage; little or no negotiation is involved. This is an extremely decentralized system, and one result is that differences in the wages of similar workers tend to be high. The variation in wages for workers of the same age, education, gender, and occupation is much greater in the United States than it is in countries with more centralized wage-setting systems.[62] That means greater earnings inequality here.

In countries like Norway and Austria, national employer associations, made up of employers in different industries, bargain with representatives of all the national unions to determine wage levels for workers in each sector of the economy. Local employers and unions are then allowed to increase (but not decrease) wages above the national level if they agree to. In countries like France and Germany, bargaining goes on between unions and employers' associations in each industry or region; the government then routinely extends these collective agreements to nonunion workers and firms in the relevant industry or region. These kinds of centralized arrangements diminish the amount of wage inequality. They do so by setting a wage floor for those at the bottom of the pay scale and a wage ceiling for those at the top of the wage scale, particularly executives.

Centralized wage setting practices are one reason why the disparity between what a typical European CEO makes and what an average European worker makes is so much less than the difference between what a typical American CEO reaps and what an average American worker earns. While top American managers might claim that their enormous compensation packages are justified by their productivity, researchers have found only a weak relation between executive compensation and productivity.[63] And, as we mentioned in the last chapter, Western Europe's economic growth outpaced ours between 1970 and 1990.[64]

One way to think about the policy choices different countries have made in the face of economic pressures during recent years is this: The Europeans have generally chosen to keep workers' real wages up, even if that means a slightly higher level of unemployment (because employers will

151

hire fewer workers at those wages). Part of that decision is a commitment to sustain the basic living standards of the long-term unemployed through government transfers and services. The United States has decided—by default—to allow real wages to drop, so that slightly more people are working but in lower-paying, less-secure, and often benefit-shorn jobs. Since no provision has been made to assist the workers in these poorer jobs, income inequality has widened more in the United States.

Both unionization rules and wage-setting practices are the result of policy choices. And these policy choices have profound effects on the amount of inequality we see in American society today. Recent statistics show that before 1974 American workers' increases in productivity were rewarded by increased wages. Since 1974, this has no longer been true. Productivity in both manufacturing and services increased by over 50 percent since then, but wages in both sectors have been essentially flat. The last twenty years' gain in productivity instead fueled gains in executive compensation and in stock prices. American workers received no greater slice of the growing pie because they had no place at the table.[65]

Public Investments:
The Case of Higher Education

Public investment decisions also shape inequality. Some investments, like clean water or public parks, improve everyone's quality of life up and down the income ladder. Other investments benefit some of us more than others. Roads that go from suburbs to downtown business areas of our large cities, for example, tend to advantage middle- and upper-class commuters more than they do central-city residents. One of the most important public investments that affects all of us, but in different ways, is public higher education.

In a crucial but not too visible manner, Americans a generation ago made a choice that moved the United States toward greater equality. From the 1950s to the 1970s, America invested enormously in higher education. In 1945 there were enough slots in postsecondary education for only one of five Americans aged eighteen to twenty-two. By 1992 the number had grown to about *four* for every five.[66] The expansion is especially impressive because it happened while baby-boomers were entering their college years. Higher education expanded enough to serve an ever greater proportion of a growing population of young people.[67]

Expansion was achieved through a generous commitment of *public* resources. Indeed, private college and university enrollments grew only slightly faster than the eligible population, while enrollments in public colleges and universities soared. States like California and New York built elaborate systems of higher education: junior colleges, state colleges, and university campuses in California, and campus after campus of the State University of New York. Other public universities increased greatly in size—the University of Michigan from 20,000 to 45,000; Ohio State from 15,000 in 1955 to 62,000 in 1975. These political choices, made largely at state and local levels, expressed Americans' optimism and belief in opportunity, the aspirations of states and cities for prestige and economic expansion, and parents' desires to assure their children's futures.

Those who believed in the link between higher education and the expansion of opportunity were right. For those fortunate enough to earn one, a four-year-college degree levels out family advantages and disadvantages in a way that increases equal access to good jobs. Among college graduates, there is *no* connection between the occupational status of their parents and their own. Children of the working class are as likely to land prestigious jobs as are children of the middle class once they have a diploma.[68] So when higher education expanded from 1960 to 1980, the intergenerational inheritance of socioeconomic status dropped dramatically. How much a father's place on the economic ladder determined what his son's or daughter's place would be was cut by half, nearly all of this decline attributable to the rise in the proportion of the workforce with college degrees.[69] (The weakening of the connection between parents' and children's statuses directly contradicts Herrnstein and Murray's argument that a genetically based intelligence is becoming more important in the modern economy. If they were right, the correlation between parents' and children's statuses should have grown stronger during those years. There are signs, however, that, with increasing tuitions and stagnating investments in higher education, the pattern of expanding opportunity is beginning to reverse.)[70]

Expansion of higher education increased equality of *opportunity* by weakening the connection between parental and child status. But overall equality of *income* depends on whether expansion of higher education keeps pace with the economy's demand for educated workers. The great development of colleges in the 1950s and 1960s increased the supply of educated workers, reducing each graduate's claim on high wages. The number of managerial and professional jobs available fell from 2.2 for each college diploma-holder in 1952 to 1.6 in the mid-1970s.[71] Better-educated workers could still bump less-educated workers from jobs farther down the ladder, but overall income equality increased.

WHOM WILL COLLEGES REJECT?

The expansion of public higher education from 1952 to 1969 (and a more modest expansion of private colleges in the 1970s and 1980s) increased equality of opportunity in America. Young people who lacked wealthy or well-educated parents were increasingly able to compete on equal terms with students from more advantaged backgrounds as seats in colleges increased. However, the picture will be sharply different in the coming decade. Between 1991 and 2001 the number of 18-to-23-year-olds will have increased by 40 percent, as the generation of the "baby boom echo" reaches college age. But in the 1990s there is neither the public will nor the fiscal resources to finance another round of expansion of higher education. Thus the proportion of young people going on to higher education will have to fall.*

One of us, Michael Hout, has examined how other modern societies handled similar enrollment squeezes. The Irish between 1970 and 1984 allocated university positions solely on the basis of achievement test scores. With the eligible population growing substantially faster than university places, this policy disadvantaged high-scoring poor and rich students equally. Italy, in a similar squeeze, kept admitting students without adding faculty or resources, so that students admitted to university increasingly found that they could not get into lectures. It took students longer to graduate, and those from better-off families were likelier to finish. Russia faced an admissions bottleneck when it expanded academic secondary schools without expanding university facilities. The shortage of places exacerbated historic inequalities, so that students from poorer backgrounds became even likelier than before to be rejected.

The United States does not provide the kind of free university education Ireland offers, so it is unlikely that the privileged and the disadvantaged will bear the brunt of the enrollment crunch equally. Instead, the United States will probably mix the Italian and Russian patterns. Students who qualify for relatively low-cost public institutions will have a harder and harder time simply finding the classes that allow them to earn degrees. The more privileged will be better able to manipulate the system and to hang on until graduation. And access to private institutions will increasingly go to those who have greater ability to pay. The result will be reduced equality of opportunity.

* Hout, "The Politics of Mobility."

After the mid-1970s, however, the supply of educated workers that colleges provided rose more slowly than the demand for them. Thus, as we first pointed out in chapter 5, the wages of college graduates rose relative to those of nongraduates. And inequality of income between those who had and those who had not graduated college increased again.[72] Today, those who do not graduate from college—and even more so, those who have a high school education or less—face bleak prospects. The earnings of college graduates are rising at a time when the earnings of high school graduates who did not attend college are falling.[73] Between 1979 and 1989 the ratio of earnings for college graduates to earnings for high school graduates who did not go to college (the "B.A. premium") rose from 1.45 to 1.65. Growth in high-tech manufacturing, health services, legal services, and the like increased the demand for college graduates. Meanwhile the decline of traditional manufacturing, bookkeeping, and commerce reduced the demand for workers with a high school education. These shifts in the kinds of jobs available in the United States economy do not account for all of the increase in earnings inequality in the 1980s, but they do account for the increased B.A. premium.[74] It is a trend, we emphasize, that reversed an earlier one and that reflects not just the market demand for workers, but also the supply provided by our decisions about investing in higher education.

In addition, American policy regarding postsecondary education is distinctive. Most of our trading partners provide students who do not go to college with more vocational training than we do. Successful systems link schools and firms. Firms can explain their labor needs to schools, and schools can draw on firms for technology and job placement.[75] The United States has given little systematic attention to vocational education, although recent research shows that vocational programs tailored to the labor market notably increase workers' earnings.[76]

Overall, then, America's investment decisions about education have had important—if complicated—effects on inequality. The expansion of higher education after World War II reduced inequality, both because it gave more youngsters who were less affluent the opportunity to attain high-paying jobs and because the growth of the supply of educated workers tended to reduce the B.A. premium. Retreats since those days have increased inequality. At the same time, the failure of the United States to invest as generously in vocational training (or in primary and secondary education; see the next chapter) has increased inequality here relative to other advanced countries where public investment in these kinds of education has been greater.

CONCLUSION: THE "FREE" MARKET AND
SOCIAL POLICY

Influential commentator George Will, responding to headlines about growing inequality in America, voiced what many Americans believe: Inequality is not bad if it results from a free and fair market.

> A society that values individualism, enterprise and a market economy is
> neither surprised nor scandalized when the unequal distribution of market-
> able skills produces large disparities in the distribution of wealth. This
> does not mean that social justice must be defined as whatever distribution
> of wealth the market produces. But it does mean that there is a presump-
> tion in favor of respecting the market's version of distributive justice. Cer-
> tainly there is today no prima facie case against the moral acceptability of
> increasingly large disparities of wealth.[77]

However, "the market's version of distributive justice" results not from
a natural market but from complex political choices, many of them hidden.
Some policies determine how unequal the starting points are of those who
enter the market's competition; other policies determine how the market
selects winners and losers. For example, African Americans in the 1950s
were prevented by private discrimination and explicit government policy
from purchasing homes and thereby lost out on subsidized loans and mortgage deductions. They were also unable to leave substantial assets to their
children. As another example, think of the businesses in industries that
receive subsidies. The market is not a neutral game that distributes just
rewards to the worthy; it is a politically constructed institution with built-
in biases.[78]

As we have shown here and in chapter 5, the enormous prosperity and
rising equality of post–World War II America resulted in part from many
government policies, some legacies of the New Deal, policies that provided
old-age security, encouraged homeownership, gave labor increased bargaining power, built massive physical infrastructure, and financed an enormous expansion of public education. Since the late 1970s, however, public
investment of these sorts has slowed and sometimes actually reversed. At
the same time, inequality has dramatically increased.

The kinds of inequalities we see reemerging in America are neither natural nor inevitable, nor do they reflect the distribution of individual talents.
Through our politics, Americans have chosen to increase equality of opportunity (expanding higher education, for example) or equality of result

(subsidies for homeownership, Medicaid, and Medicare, for instance), but to do so to a far more limited extent than citizens in other nations have chosen. We extend support to fewer of our citizens, largely the elderly and the middle class; and we extend less support. For example, we provide medical insurance for some residents; most nations provide medical care for all. We provide a tax deduction for children of taxpayers; most nations provide family allowances. Americans have also made choices that increased inequality, such as the tax changes of the 1980s and the rules on unionization we have accepted. We have structured many programs to help the well-off more than the less well-off, such as the subsidies for homeownership and medical insurance.

What all this implies is that the inequality we see today in America is in great measure a result of policy decisions Americans have made—or chosen not to make. Generally, we have chosen to do far less to equalize life conditions than have other Western people. We have chosen to reduce the inequality between the middle class and the upper class somewhat, but to do far less to reduce the gap between the lower class and other Americans—with the notable exception of older people. And in the last couple of decades, our choices have moved us farther from equality. Some criticize these choices; others, like George Will, may applaud them. Either way, Americans constructed the inequality we have.

Enriching Intelligence:
More Policy Choices

I F WE HAVE done our job well, readers of this book should by now appreciate that the explanation for inequality lies in the design of society, not in the minds or genes of individuals. The rewards that contestants gain in the race for success are determined by the rules of the race, not the personal traits of the racers. Even who wins or loses the race is determined more by the "nutrition" and "training" they receive than by their "natural speed." In the last few chapters, we have directed attention away from individuals to social systems and shown that inequality and its reduction are to be found in structures of competition and distribution. The leverage for expanding opportunity and moderating inequality lies in policies that deal with those structures.

In this chapter, we argue that social policy can also improve the cognitive skills—or, for shorthand, the intelligence—of individuals. This is a worthwhile goal, because a more cognitively skilled population would be a more productive one, a wider distribution of skills would expand opportunity, and cognitive skills are valuable in their own right. Those who believe that intelligence is genetically determined or fixed in infancy claim that nothing can be done to increase such skills; intelligence is immutable. That is why, they argue, intervention programs are a waste of taxpayer money. However, the pessimists are wrong. Evidence shows that intelligence can be changed. Indeed, it is shaped every day by institutions all around us. This is true if we define intelligence in its broadest sense, as it should be. But even if we define it narrowly, as school skills (see chapter 2), it is still changeable. To improve skills, we need to improve the *social environment*.

The debate about the mutability of intelligence has been misfocused; it has lost the forest for the trees. Both the pessimists and the optimists have argued about whether special interventions, such as Head Start and remedial instruction, can raise test scores. But the real leverage lies not in such episodic events; it is, instead, in the *continual, systematic, everyday* ways society forms intelligence. We will therefore forgo debating those intervention programs, only noting for the record that many scholars have per-

158

suasively defended their effectiveness (see, e.g., Dickens, et al., chap. 3). Instead, we will look at four examples of how society pervasively shapes cognitive skills: schools, summer vacations, classroom tracking, and the structure of jobs.

DO SCHOOLS MATTER?

In 1964 Congress mandated a major study of educational inequality in America, asking whether disparities in school resources accounted for differences in students' test scores. Some members of Congress hoped the findings would arouse the nation to battle ignorance by equalizing educational resources across schools. Alas, the best-laid plans often fail. After painstaking analysis and reanalysis, the late University of Chicago sociologist James S. Coleman was unable to find that school resources had much of an effect on student achievement. Neither the number of books in the library, nor the number of credentials on the teachers' résumés, nor the size of students' classes, nor the length of the school day seemed to affect achievement scores. Coleman concluded that family environments, rather than schools, determine learning.

The report set off a firestorm. How could schools be irrelevant, when parents move to expensive suburbs just to enroll their children in highly regarded schools? Did the Coleman Report really lay the blame for students' failures at the feet of the parents? And, finally, what good is public policy if resources in schools do not matter?

Sociologists today understand how schools shape inequality in learning much better than they did in the 1960s. Yet some commentators ignore the insights gained in the nearly thirty years since the original Coleman report. For example, Herrnstein and Murray cite the 1966 Coleman Report as demonstrating the futility of attempting to raise intelligence. But they ignore the many studies critical of the original 1966 report, including Coleman's own subsequent work.[1] The primary value of the 1966 Coleman report today is as a historical demonstration of how a mis-asked question can lead to a mistaken answer.

Correcting limitations of the original study and applying better methods, researchers have found that variations among schools *are* important. One key limitation of the 1966 report is that it assumed that every student in a given school has access to the same resources, the same books, and the same teachers. But we cannot make this assumption. Having many books in the library can increase the achievement only of those students exposed to the books; quality teachers help only those students who enroll in their classes. Coleman's approach did not take into account individual students' exposure to the resources of the school when looking at their achievement scores, and thus he underestimated the effects of those resources.

Researchers have devised better understandings of how educational achievement is produced *inside* schools. They have also developed better understandings of how children learn, which has led them to ask sharper research questions, such as what affects students' *rates* of learning. And researchers have expanded their analysis beyond simply comparing schools, looking, for example, at how much students learn during school compared with when they are out of school. These kinds of studies show that schools and school resources *do* matter.

Understanding How People Learn

One important post-Coleman advance was a deeper understanding of how people learn. Consider learning to ride a bike. When one first begins, one knows nothing. Learning at the beginning is likely to proceed very slowly. As one begins to feel comfortable with the task, the *rate of learning* speeds up. Learning does not, however, increase forever; eventually it reaches a plateau. There is always more to learn, and some bikers become extremely skilled, but after some plateau is reached additional learning usually proceeds slowly. Figure 7.1 demonstrates the classic "learning curve."

The slope of the curve indicates the speed of learning: the steeper the slope, the faster the learner learns. Different people have different learning curves. When we say that someone is "bright," we often mean that he or she is a fast learner in just this sense. If we can measure people's skills frequently enough, we can measure each one's learning curve. We can then compare those curves to see what circumstances accelerate or slow down individual learning. *Speed of knowledge growth* is the very issue in which one should be interested if one is interested in intelligence. For example, Herrnstein and Murray claim that intelligence makes some people more accurate, rapid, and efficient in solving problems and attaining success. If so, then the speed with which a person's knowledge grows could be defined as their intelligence or, at the very least, as a good proxy for their intelligence. Researchers studying learning curves find powerful evidence that *schools are the primary determinants of speed of learning*. This turns the old Coleman findings upside down.

Anthony Bryk and Stephan Raudenbush, of the University of Chicago and Michigan State University, respectively, studied the learning curves of

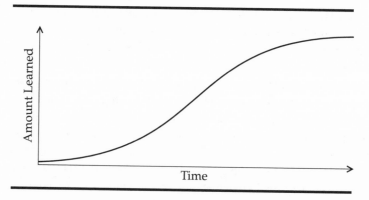

7.1. An S-shaped Learning Curve

618 students in 86 different schools as they moved from first grade to third grade. They investigated students' learning curves in mathematics and in reading. This distinction is important because the only place students are explicitly taught math is in school, while language skills are taught from birth and through most social interactions children have. When earlier researchers found that variation among schools did not account for much of the variation in students' math scores but mattered more for language skills, they concluded that schools could not be important because if they were, schools should matter especially for math.[2]

Bryk and Raudenbush found what Coleman found when they asked the same question Coleman asked: How much variation in mathematics and reading *achievement* is due to the schools the children attend? They found that only 14 percent of the variation in mathematics achievement and 31 percent of the variation in reading achievement was attributable to differences between schools. Unlike Coleman, however, Bryk and Raudenbush had data that allowed them to investigate learning curves. They came to a startling conclusion. Over 82 percent of the variation in mathematics learning *curves* occurred between schools; that is, 82 percent of the variation in how quickly students learned math was attributable to school differences. Nearly 44 percent of the variation in learning curves for reading also occurred between schools. The finding is what one would expect if schools matter: schools are more important for math than for language skills.

Thus, at its core, the answer of the Coleman report was incorrect. If the average test scores for students in school A and in school B are pretty much the same, and if scores seem to be accounted for by the students' family backgrounds, then it appears that schools—their funding, structure, leadership, resources—do not matter. Yet, more complete analyses show that schools matter very much. Schools matter because differences *within* schools in how students are taught affect learning. (We will look at one instance, tracking, below.) And schools matter because they influence *rates* of learning.

That schools matter may seem an unexceptional conclusion; after all, many American parents struggle to put their children into "good" schools. But the conventional academic wisdom for years was that the payoff for those struggles was not academic. Cognitive skills seemed to be affected only by family background—genetic background for many psychometricians, social background for sociologists. (Perhaps the payoff lay in the social or "networking" advantages of good schools.) The conventional political wisdom in recent years has been that investing in schools, particularly for the less advantaged, was wasteful. *The Bell Curve* seems to certify these views, but here again it misleads us. Schools appear to be the primary determinant of children's *intellectual* advancement.

SUMMER VACATIONS

Other researchers also interested in the question of whether and how schools affect learning and test scores have taken advantage of a routine feature of American life, the summer vacation. It is the major reason most American children receive only 180 days of instruction per year. Researchers compared how much test scores change during the school year with how much they change during the summer in order to see what effect being in school has on cognitive skills.

The logic of summer learning research is this: We can see how schools affect development by comparing how much children learn when they are in school with how much they learn when they are out of school. To do that, researchers compare changes in test scores that occur during the school year, from fall to spring, to changes in test scores that occur during summer vacation, from spring to fall. Researchers have found—and it is no surprise except to those who doubt that intelligence is mutable—that children increase their intellectual performance much more during a month in school than during a month of summer vacation. Furthermore, what happens to children's skills during the summer depends a lot on their summer *environments*. Children from disadvantaged circumstances tend to stagnate or even fall back intellectually during the summer, while those in better-off circumstances make some gains during the vacation. It looks as if middle-class parents provide their children with experiences, such as camp or travel or lessons, that add to cognitive growth and higher scores, while poorer parents do not. Just like going to school, having summer activities increases test scores.

In studies separated by some fifteen years, at different stages of the alleged destruction of standards in schools, Barbara Heyns, then at the University of California, Berkeley, and Doris Entwistle and Karl Alexander at The Johns Hopkins University found that all students learn at a faster rate during the school year than they do during the summer. Some students' rates of learning are so slow during the summer that they fall behind their peers. Poverty is the important determinant of which students tend to fall behind during the summer. The researchers also found that the disadvantages of poverty are compensated for during the school year; that is, poor children begin to catch up to middle-class students *while school is in session*. In reviewing the literature on summer learning, Heyns found that the entire black-white gap in reading achievement is due to differences that emerge during summer vacations. These findings imply that schools are very effective because they overcome the disadvantaged backgrounds of students. But schools cannot sustain their effectiveness in improving the skills of the disadvantaged because schools are in session only nine months of the year, whereas family disadvantages operate during winter, spring, summer, and fall.[3]

Many weaker students might gain academically if schools were to become year-round. Herrnstein and Murray concede in passing that year-round schooling might reduce inequality but then dismiss the point by arguing that it is politically inviable.[4] The tradition of summer vacation will likely die hard. However, it is a matter of policy; there is no natural imperative for school buildings to close during June, July, and August, nor for children to take a three-month vacation from formal schooling. (It is a leftover from the days when children were needed to help on the farm.) Respecting the summer vacation tradition is also to choose lower levels of intellectual skills and a higher level of inequality. We could reduce disparities in cognitive skills if we chose to.

TRACKING

In the wake of the Coleman report, many researchers turned their attention to what might be called "schools within schools," to tracking. This is the process of assigning some students to more rigorous college preparatory courses and other students to less rigorous preparation for immediate work careers. Essentially, it creates different learning environments within one school.

Higher–scoring students end up in college tracks, but at each level of test score, students have a better chance the higher their class background.

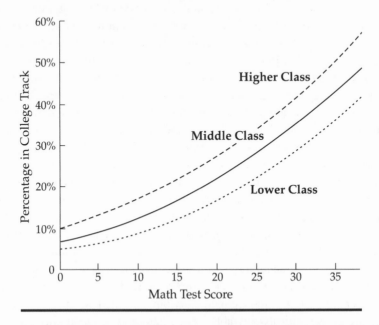

7.2. Probability that Students Were in College Track by Math Test Score and Social Class (*Source*: Adapted from Gamoran and Mare, "Secondary School Tracking")

Researchers have focused on two questions: (1) Are track assignments fair? (2) Do track assignments matter? By fairness, researchers typically mean that equally well-prepared students—as measured by previous achievement—of whatever ethnicity, gender, or class have an equal chance of being assigned to the college-preparatory track. The vast majority of analyses show that parents' social class strongly influences who is assigned to the college track.[5] This is one way parental background sustains inequality, by shaping children's academic development.

Figure 7.2 shows the estimated probability of entering the college track for white boys of varying mathematics achievement. The horizontal axis shows tenth-grade students' scores on a math screening test, and the vertical axis shows the chances that a student was in a college track. The heavy middle line indicates the probability that a student from a family of average social class would be in a college track, given his math score. The lines above and below that one represent the chances that students from better-off homes and from worse-off homes, respectively, would be in a college track.[6] At each level of test score, students from better-off families had a greater chance of being placed in a college track than did students of poorer class backgrounds. At the highest level of measured mathematics achievement, economically disadvantaged students suffered a fifteen-point penalty in the chances of being in the college track. (Compare the highest and lowest lines at the far right of the graph.) The lower-class student who achieved at the highest possible level had about a 42 percent chance of being in a college track, while the higher-class student with the identical score had about a 57 percent chance. Put another way, students who miss one-quarter of the mathematics questions but come from advantaged backgrounds had as much chance of being in the college track as the impoverished student who answered *every* question correctly. (These comparisons are fairly pure because the possibly confounding effects of the schools' structure and composition, as well the effects of individual achievement in other subjects, have been statistically held constant.)

The class bias in track placement, large as it is, would not matter if being tracked did not affect achievement and college entry. But students in college tracks not only learn more facts and learn different facts, they also develop a different relationship to knowledge (see box, p. 166).

Many studies show that students in the college track learn more than do students in the noncollege track.[7] One example will show how tracking alters test performance. In an especially well-executed study, Alan Kerckhoff of Duke University used data on all children born in England, Scotland, Northern Ireland, and Wales in the week of March 3–9, 1958, to compare students placed in high tracks, students in low tracks, and students in untracked schools. This design allowed Kerckhoff to control for many of the factors we already know might be related to achievement, such as students' social class, aspirations, and prior educational experience. Kerckhoff had information on the students at age seven, age eleven, and age sixteen. He found that high-track students gained *more* academically than students in the untracked schools, and that low-track students gained *less* than students in the untracked schools. This pattern widened preexisting differences. Thus, after students were assigned to different tracks, their achievement levels diverged; tracking exacerbated inequality.

An obvious rejoinder to the Kerckhoff study is that the students assigned to high tracks were more intelligent at the outset. Thus, one would expect high-track students to learn so much more over time that the gap between them and low-track students would grow. But Kerckhoff compared

---

SAME BOOK, DIFFERENT LESSON

Researchers have shown that many school practices undermine instruction in lower tracks. For example, high-track students are often given demanding material, reading Shakespeare, Faulkner, and Walker, while their peers in low-track classes read "young adult fiction." But even when low-track students are given the same books, the uses to which the material is put result in vastly different learning experiences. A high-track class might write and produce a play that captures the essence of *The Color Purple*, or, using *Julius Caesar* as a point of departure, argue over whether and how one can justify murder. Students in the lower tracks are more likely to fill out worksheets that have a rote structure of question and response, and where all of the "correct" responses are to be found in the book. The high-track students are learning to treat knowledge as something one uses and can even produce for one's own ends; in contrast, the low-track students are taught to see knowledge as something outside of themselves, beyond their power, that stands over and against them.*

  *Cookson and Persell, *Preparing for Power*; Anyon, "Social Class and School Knowledge"; Bernstein, *Class, Codes, and Control*.

---

students who had the same initial achievement level and yet had been assigned to different tracks. He found that students assigned to the high track gained more than otherwise similar students assigned to lower tracks. Thus, if, as many claim, these tests reflect intelligence, Kerckhoff's study shows how tracking selectively nurtures or neglects existing intelligence. Those whose intelligence is nurtured gain; those whose intelligence is ignored lose.

Kerckhoff studied Great Britain; the same kind of finding appears in the United States.[8] The data show that tracking has powerful effects on cognitive growth and that students excluded from demanding instruction learn less than they would have otherwise. Because gains for some students are offset by losses for others, the average level of achievement of all the students in tracked schools is not increased. Yet inequality is increased.

Tracking exists in elementary and junior high as well as secondary school. Even though the systems of tracking are far more varied in the lower schools, analysts find the same pattern for younger students we have discussed in the case of older students: powerful effects of social class on student placement, and powerful effects of track location on students' success that exacerbate preexisting cognitive differences.[9] There is also a broader sense in which American schools are tracked. The great decentralization and local control of schools in the United States—much more than is common in other Western nations—means that entire schools and even school districts are, in essence, on different tracks.[10]

In sum, studies of tracking suggest that we *do* change children's academic intelligence all the time. The entire process of tracking is designed to do just that. Most Americans may have good reasons for adopting tracking; it seems to advantage middle- and upper-class students. Less well-off students, however, are shunted disproportionately to lower tracks. Inequality in learning, although not the average level of learning, is increased through tracking. By these practices schools demonstrate the pliability of cognitive skills as well as the powerful effect social factors have on the success of individuals. Policies alter intelligence.

ADULT DEVELOPMENT AND JOBS

So far we have looked at how schools affect children's intellectual abilities. That makes sense, because schools are the major institution we have designed precisely to expand people's cognitive abilities. But intellectual development does not stop with school nor end in people's childhoods. Older notions of intelligence led psychologists to assume that, after about age twenty, people stopped growing mentally and probably got dumber. Intelligence is, by this view, fixed early in life. Psychometricians believed that cognitive skills deteriorated after young adulthood because, as we discussed in chapters 2 and 3, their measures of intelligence are measures of what people learn in school and, of course, as people age they forget much of that. But modern psychologists have expanded their notions of intelligence to include a variety of abilities, including practical problem-solving. The result is that they now believe that cognitive skills and practical intelligence typically increase into the forties, stay level for ten or twenty more years, and then tend to decline after sixty. Even then, however, not all skills deteriorate. Moreover, gerontologists have found that even the skills that do tend to weaken can be restored with training. Just five hours of training can, for example, substantially improve older people's inductive reasoning.[11]

Not everyone grows intellectually as they move toward middle age, nor does everyone decline in old age. What matters are the situations that people face and the experiences that they have; some environments stimulate cognitive development and some stifle it. Getting more education, staying culturally active, even being married for a long time to an intelligent spouse—these are the kinds of activities that spur the growth of cognitive abilities or at least slow their deterioration.[12]

One crucial experience that affects adults' mental development in these ways is their *jobs*. Social psychologists Melvin Kohn and Carmi Schooler, working out of the National Institute of Mental Health, have conducted decades-long research across three continents on how different kinds of jobs affect people psychologically. Kohn and Schooler have measured how much self-direction jobs require. The more complex the job, the less routine it is, and the less supervision it entails, the more self-directed the job is. Among the consequences of self-direction Kohn and Schooler have studied is what they call "intellectual flexibility"—in effect, intelligence. And they have found that "job conditions that promote occupational self-direction increase . . . intellectual flexibility, whereas jobs that limit occupational self-direction decrease . . . intellectual flexibility."[13]

A quick and obvious objection might be that intelligent people seek out and get self-directing jobs; that is what explains the correlation between having a demanding job and having good cognitive skills. That is true, and much of the causal connection between job and psychology runs that way. But, using various techniques and data sets, Kohn and Schooler have been able to show that the causal effects run both ways. Having a complex job that is intellectually demanding itself increases people's cognitive functioning, whatever their prior intelligence.

Upon reflection—and upon other research—it is clear why that would be so. Spending eight hours a day, five days a week, fifty weeks a year in a position that forces one to analyze problems, to calculate, to strategize, often to persuade others, and to do these tasks independently exercises the mind, just as a rigorous calisthenics regime exercises the body. Spending those same two thousand hours a year in a job that only asks one to repeat the same routine actions and to follow orders is to the mind what spending hours each day as a "couch potato" is to the body.

The psychological consequences of job structure, it turns out, go beyond the jobholders themselves to their children. Social scientists have long known that middle-class parents tend to teach their children independence and self-direction, while working-class parents are more likely to emphasize conformity and being obedient. Melvin Kohn showed that one reason for this difference is that middle-class adults usually have had more secondary and higher education, a training that stresses independent thinking. Kohn also showed that another reason for the class difference is the job experience. "Men of higher class position, who have the opportunity to be self-directed in their work, want to be self-directed off the job, too, and come to think self-direction possible. Men of lower class position, who do not have the opportunity for self-direction in work, come to regard it a matter of necessity to conform to authority, both on and off the job. The job does mold the man—it can either enlarge his horizons or narrow them."[14]

These different orientations then carry over into how adults parent—when and how they discipline their children, for example. Blue-collar workers are more likely to expect their children to do what the parent says because the parent says so; white-collar workers are more likely to expect their children to be able to explain *why* the parent says what he or she says. This training, for obedience or for independence, in turn prepares children differently for their own work lives.

> The conformist values and orientation of lower- and working-class parents, with their emphases on externals and consequences, often are inappropriate for training children to deal with the problems of middle-class and professional life. . . . [C]onformity is inadequate for meeting new situations, solving new problems—in short, for dealing with change. . . . [T]he self-directed orientation of the middle and upper classes . . . is well adapted to meeting the new and the problematic. At its best, it teaches children to develop their analytic and empathic abilities. These are the essentials for handling responsibility, initiating change rather than merely reacting to it. Without such skills, horizons are severely restricted.[15]

The job is one setting that, well into adulthood, shapes intelligence and other psychological dispositions that, in turn, affect people's life outcomes. While most jobs would seem to be outside the purview of governmental policies, most are under the control of large private organizations. Evidence shows that, within limits, employers who enrich the intellectual demands of the jobs they provide also improve the cognitive skills of their workers.

The broader point is that cognitive skills keep changing over the life course and are changed by experience. Policy can intervene here by, for example, increasing older people's opportunities for intellectual stimulation.

CONCLUSION: THE ONLY CONSTANT IS CHANGE

Despite the rhetoric of the intelligence-never-changes school, we should not be surprised to learn that cognitive skills do change over the life course.

Indeed, researchers have long known that those skills change so much that early intelligence scores are weak predictors of later intelligence test scores. The correlation between measured intelligence at age six and age seven is .86, but the correlation between measured intelligence at age six and age fifteen is only .69. Were intelligence immutable, we would expect the correlation to be close to 1, or at least we would expect the correlation not to decline as the gap in ages increases. But children's ranks on IQ tests change so much that their scores at age six account for less than half of the variation in those scores at age fifteen.[16]

Why should such change in IQ occur? Because, as we have noted, children experience different school environments, and those environments play an important role in how fast they learn. Children, locked out of school in the summer, cope variously well or poorly with the absence of formal cognitive training. Children are assigned to different tracks, and some of those tracks lead to heightened self-efficacy and greater exposure to knowledge and analysis, while other tracks lead to lowered self-efficacy and reduced exposure. But children are not alone in requiring a nurturing and challenging environment in which to grow; even adults lose their edge faster when the environments in which they work and live do not test their capabilities. (Other activities, too, such as television-watching, also shape cognitive abilities.)[17]

A common theme here is that challenges and mental exercise expand intellectual abilities. People, be they children in demanding schools or adults in demanding jobs, grow to meet those challenges. When their skills are tested, people usually strengthen those skills; when not, those skills tend to atrophy. Contrary to the fatalists who claim that intelligence is determined at birth or at least fixed by adolescence, cognitive ability—even narrow, academic intelligence—is quite changeable.

American policy choices, whether obvious like decisions about tracking and school funding or more subtle like decisions about vacations and job structure, shape American intelligence. To the extent that academic intelligence, as measured by grades, SATs, and the like, is used to sort out people for advancement in school and the economy, then these same policies currently shape American inequality. Because we already shape cognitive skills, we can reshape inequality.

Race, Ethnicity, and Intelligence

N<small>O CHAPTER</small> in *The Bell Curve* received more attention than the one claiming that African Americans and Latino Americans are inherently less intelligent than white Americans. This attention is not surprising, because Americans look at virtually every issue—crime, poverty, politics, housing, sex, and even sports—through the prism of race. Herrnstein and Murray raised this sensitive topic even though their essential argument, that *individual* differences in intelligence explain social inequality, neither stands nor falls on the question of *group* differences in intelligence. Although the issue of race is irrelevant to their main argument (it is there to justify a critique of affirmative action), their book added to the American obsession with race. We, however, needed to treat racial differences in this book, because race has been and still is the great chasm in our society. We cannot understand inequality in America without addressing the roots of racial inequality and the controversy over intelligence.

For readers who may have turned first to this chapter precisely because it deals with race, we reiterate the basic findings of our earlier chapters: *The Bell Curve* is mistaken. The test of intelligence Herrnstein and Murray use, and most others, too, really measure how much academic instruction people have had, not their inherent abilities. Even if such tests did measure differences in native intelligence, those differences do *not* explain very much about inequality among individuals in America; individuals' social environments explain more. More significantly, the distribution of individual intelligence has little to do with the extent of inequality in a society; patterns of inequality are produced by the economic and social structures of the nation and era. Both the conditions that help or impede individuals' race to succeed and the system of inequality within which those individuals compete are heavily governed by specific social policies. Thus, policy choices have shaped the kind of class inequality we have. Policy choices, over the long course of American history, have shaped the kind of racial inequality we have, as well.

THE ARGUMENT ABOUT RACE AND INTELLIGENCE:
A PREVIEW

African Americans and Latino Americans in the United States tend to be economically worse off and to suffer from more social problems than do whites. Blacks and Latinos[1] also tend to score lower than do whites on standardized tests. For those who believe that inequality is "natural," the second pattern clearly explains the first: Ethnic groups are socially unequal because they are intellectually unequal. But for those who understand that societies construct the inequalities they have, as has been demonstrated in this book, the reverse is true: *Groups score unequally on tests because they are unequal in society.*

Consider this situation: Members of a minority, many of whom were brought to the country as slave labor, are at the bottom of the social ladder. They do the dirty work, when they have work. The rest of the society considers them violent and stupid and discriminates against them. Over the years, tension between minority and majority has occasionally broken out in deadly riots. In the past, minority children were compelled to go to segregated schools and did poorly academically. Even now, minority children drop out of school relatively early and often get into trouble with the law. Schools with many minority children are seen as problem-ridden, so majority parents sometimes move out of the school district or send their children to private schools. And, as might be expected, the minority children do worse on standardized tests than majority children do. What is this minority?

Koreans in Japan.

Koreans, who are of the same "racial" stock as Japanese and who in the United States do about as well academically as Americans of Japanese origin (that is, above average), are distinctively "dumb" in Japan. The explanation cannot be racial, nor even cultural in any simple way. The explanation is that Koreans, whose nation was a colony of Japan for about a half-century, have formed a lower-caste group in Japan.[2]

Consider another case: Immigrants flood into the United States to take low-wage jobs. They are "swarthier" and more "primitive" than most Americans; they seem unwilling or unable to assimilate; they are suspected of criminal behavior; and they threaten native workers' jobs. Together with other newcomers, these immigrants provoke a backlash against easy immigration into the country. Objective test data show that the immigrants are low on intelligence; and their children do poorly in school. Who does this describe?

Polish Jews in the 1900s and 1910s.[3]

These same Polish Jews, whose descendants now do well both economically and on tests, were scorned when they came. Many scholars of the day believed that the Jews, along with Italians and other "dark" types, were dim-witted.[4] Can race or genes explain what happened in the intervening seventy or so years? No, but social location can. As a peripheral and subservient group in the early twentieth century, Polish Jews were failures, even in school. As they came to be accepted in American society, their position and their "intelligence" rose.

Our argument, which we will flesh out later in the chapter, is that *a racial or ethnic group's position in society determines its measured intelligence rather than vice versa.* Some ethnic groups find themselves in inferior positions through conquest or capture (e.g., the Irish in Great Britain, Maori in New Zealand, Africans, Mexicans, and Indians in the United States). These are the groups that suffer the most drastic and lasting effects of subordination. Other minority groups enter a society through immigration, often made under economic duress (e.g., Italians in the United States, Eastern-origin Jews in Israel, Turks in Germany). These groups typically move closer to equality faster.[5] In either case, subordination leads to low performance in three ways.

First, individual members of the subordinate groups and the families they grow up in suffer *socioeconomic deprivation.* Low income, ill health, poor parental education, and the like reduce test performance. This process is a familiar one and the subject of much research. When Herrnstein and Murray say that socioeconomic differences between individual blacks and whites do not explain the black-white gap in test scores, they are asserting that this first process—deprivation based on differences in economic conditions—is insufficient to account for the race differences. This is the only process they examine. The second process, however, cannot be understood individualistically.

Subordinate groups typically experience *segregation.* Ethnic ghettoes concentrate disadvantages and accentuate them. (The term "ghetto" originally described Jewish quarters in European cities. Ironically, the immigrants to America from those ghettoes were viewed in a similar light generations ago as today's ghetto residents are viewed.) Segregation also exposes children who would otherwise do well to the problems and the culture of the disadvantaged.

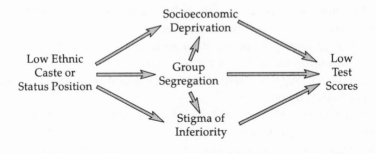

How inferior ethnic caste position leads to low test scores.

8.1. A Model of How Low Ethnic Position Causes Low Test Scores

The third process is cultural and psychological: Young members of subordinate groups understand that they carry a *stigma of inferiority* based on the wider society's perception of them. The young people respond in various ways to that identity. Some become anxious, fatalistic, and resigned; others reject the wider culture's expectations and standards, adopting an oppositional stance. Either of these reactions to stigmatizing images— resignation or rebellion—brings down average school and test performance. Youth from advantaged ethnic groups have mirror-image experiences of all these. They typically benefit from more affluent family circumstances, from having friends and neighbors who do well and who can help them to do well, and from the confidence that they are destined to succeed. Figure 8.1 outlines the argument.

We do not claim that this explanation applies equally well to all ethnic groups in all societies; it is important to understand that groups' particular histories matter. But we do claim that it applies to most cases and that it applies particularly to groups that have been severely suppressed, such as American Indians, Mexican Americans, and African Americans.

Understood this way, groups' test scores are not the *beginning* of an explanation for inequality but the *end* of one. The beginning is history. That is why the next section reviews the history that created a racial caste system in the United States. In the subsequent section, we describe the continuing legacy of this caste system in our time. In the final section, we explain why blacks and Latinos tend to score below whites on tests of academic skills. We show that these lower scores fit a worldwide pattern of ethnic group differences that has nothing to do with racial intelligence. Instead, the American case, like others around the globe, is explained by the three consequences of caste: deprivation, segregation, and stigma. *Minorities score lower because they are lower caste.*

ETHNIC HISTORY AND ETHNIC INEQUALITY IN AMERICA

Herrnstein and Murray, like many others, seem to believe that the history of slavery, segregation, and discrimination in America, a history they acknowledge but then forget, ended sometime in the 1960s.[6] The slate was wiped clean; everyone then started out even. Any group inequalities in outcomes since the 1960s, therefore, can only be explained as the result of group inequalities in natural talent. But life is not so simple. Even if discrimination had ended in the 1960s—and it certainly did not (see next section)—the weight of history is oppressively heavy. Three decades of moderate reform have narrowed some of the gaps, but Americans were naive to think that it would quickly erase three centuries of a rigid caste system.

A *caste system* ranks groups economically, politically, and socially. The privileges of those ranks are usually enforced by law and policy. People belong to the ranked groups by virtue of their birth; the groups are in that sense ethnic or racial. The system is justified by a set of beliefs, often shared by the lower-ranked as well as higher-ranked groups, about the intellectual and moral superiority of the higher-ranked one.[7] The term "caste" is drawn from India, but it describes perfectly the American race system until the 1960s.

Although most non-English people who came to America faced discrimination and exploitation, the experiences of a few were distinctly different. What separates the historical experience of Africans from that of other groups is that they did not come to the United States to find work but were instead impressed into labor. They, like American Indians, form what Berkeley anthropologist John Ogbu has labeled an "involuntary minority" and are distinct from voluntary immigrant groups such as the Irish, Jews, and Chinese.[8] The African case is well-known: Europeans purchased captive Africans and sold them in America as slaves.

The Mexican American case is more complex; it fits neither Ogbu cate-

175

gory neatly. (We focus in particular on Mexican Americans because the Mexican experience in the United States is closer to that of blacks; it is distinct from that of most other Latino groups, such as Central Americans and Cubans; and Mexicans are the largest Latino population. The case of Puerto Ricans, one-sixth the population of Mexican Americans, however, bears many similarities to that of Mexican Americans and African Americans.) After the Mexican-American War of 1848, when the United States conquered Texas, California, and the Southwest, many Mexicans were caught under the control of the enemy they had just fought. The Americans treated the Mexicans both as beaten foes and as exploitable labor.[9] Later immigrants to the United States, although voluntary, were absorbed into a conquered group. Moreover, those immigrants were treated by Anglos, as we shall see later, as members of a racial caste. Thus, the Mexican case was different in origin but similar in outcome to that of involuntary migrants, such as Africans.

Although educated Americans ought not to need reminding of our cruel history of slavery, a few salient points are worth noting. First, American slavery was, historically, one of the more rigid versions of slavery in the world. Africans were not only taken thousands of miles from their homelands but were also shorn of much of their culture. Americans built an especially high wall between the status of slave and that of freeman and built it higher over time. In most states any drop of "black blood" condemned one and one's descendants forever to slavery, unless formally emancipated. The race-and-slavery borderline was ferociously guarded with respect to sexuality. A second distinct feature of American slavery was the elaborate ideology southerners developed to reconcile slavery with a democratic society. They began with religious justifications, that blacks were descendants of Noah's dark son Ham, whose transgressions condemned his descendants forever to be servants. They ended with "scientific racism," scholarly arguments that Africans' biological "nature" made them fit only to be servants. Third, American slavery ended later than did slavery elsewhere in the advanced world.

These distinctive features of American slavery contributed to the enormous weight of disadvantage passed on to the descendants of slaves. The formidable edifice of slavery was, of course, not natural; it was policy. Americans decided in 1789 to allow slave importation until 1808 and for over fifty more years after that decided to permit slavery to expand. We continue to live with the consequences of those choices.

Mexicans were not slaves, but in the Southwest they formed a caste as distinct from Anglos as blacks were from whites. Anglo Americans

176

Intelligence as Rationalization

When groups find themselves in situations of inequality, the dominant one generally develops a justification for its advantages. A theory of innate intelligence is a perfect example of such a justifying ideology. Americans used this theory to explain why Africans and Mexicans—and others, too, such as the Irish and American Indians*—were subordinate. Although both Africans and Mexicans had established complex societies in their native lands, Anglo Americans believed them to be intellectually inferior. Thomas Jefferson provides an example of such an interpretation:

> Comparing [blacks] by their faculties of memory, reason, and imagination, it appears to me, that in memory they are equal to the whites; in reason much inferior, as I think one could scarcely be found capable of tracing and comprehending the investigations of Euclid; and that in imagination they are dull, tasteless, and anomalous. . . . Most of them indeed have been confined to tillage, to their own homes, and their own society: yet many have been so situated, they might have availed themselves of the conversation of their masters; many have been brought up to the handicraft arts, and from that circumstance have always been associated with the whites. Some have been liberally educated, and all have lived in countries where the arts and sciences are cultivated to a considerable degree, and have had before their eyes samples of the best works from abroad. . . . But never yet could I find that a black had uttered a thought above the level of plain narration; never see even an elementary trait of painting or sculpture.**

Jefferson struggled with the contradiction between his egalitarian principles and his slave practice. Although Jefferson thought the institution of slavery was tyrannical, he understood certain practical considerations. Slavery provided the labor needed to sustain the plantation economy. Abolishing it would create economic hardship for the general citizenry and potential political chaos. A theory of innate intellectual inferiority helped reduce the dissonance. Logic could be sacrificed in this effort by implying, for example, that by working near his master a slave might absorb a bit of

---

* On the Irish, see Roediger, *The Wages of Whiteness*. On such arguments regarding American Indians, see Rogin, *Fathers and Children*, p. 34.

** Jordan, *White over Black*. Jordan draws his quotation from Jefferson's *Notes on Virginia*.

Euclid. Jefferson also dealt with the dissonance by suggesting that the question of innate inferiority be left to future scientific observation, a solution remarkably similar to that disingenuously advanced by Herrnstein and Murray.***

*** Ibid., pp. 438–39; Herrnstein and Murray, *The Bell Curve*, p. 317. There are innumerable references in *The Bell Curve* concerning the need for future scientific research to determine the real impact of genetics on the cognitive abilities of different races. Thus, like Jefferson, they hedge on the ability to account scientifically for real genetic differences in cognitive ability and leave it for future scientific inquiry.

considered Mexicans both ethnically different and former enemies but also saw in them a source of badly needed labor. Anglos adopted the *patrón-peón* system to use that labor. In this system, the *peón* ranchworker relied on a landowning *patrón* to provide him with a job, home, security for his family and his old age, and even religious instruction of his children (through godparenthood). For these favors, the *peón* reciprocated with total loyalty throughout his lifetime.[10] This quasi-feudal structure, which operated not unlike slave plantations in the South, provided a social order useful for economic development. Anglos did not need to segregate Mexicans during the early years because the *patrón-peón* system was sufficient to maintain caste lines and guarantee Anglo control. Anglos did not impose the most malignant forms of racial ideology until the 1920s when they adopted the same sort of Jim Crow system that African Americans then faced.[11]

Nineteenth-century claims about differences in intelligence provided a rationalization that was comforting to whites but a poor explanation for the disadvantaged positions of African Americans and Mexican Americans.[12] Physical intimidation, legal codes, and social custom are sufficient to account for the latter's lower caste position. Apologists invoked intelligence to justify the caste order only after whites had physically subordinated blacks and Mexicans.[13]

The slave and feudal systems that had been in place gave way at the end of the nineteenth and beginning of the twentieth centuries to a rigid and malevolent caste system based on tenant farming and segregation. Understanding this system is critical for understanding race relations today, because it persisted until recently. Many white Americans dismiss the oppression of blacks by saying that slavery ended over a century ago. True, but virtual serfdom lasted much longer. This serfdom was only yesterday for blacks and Mexican Americans.

White southerners established the Jim Crow system over the quarter-century following Reconstruction's end in 1877, when federal authority withdrew from the former Confederacy. Jim Crow laws and vigilante repression such as Ku Klux Klan lynchings were designed—with neither embarrassment nor euphemism—to sustain white supremacy after the Emancipation Proclamation. Those laws established segregation, eliminated civil liberties, withdrew the right to vote, denied educational and employment opportunities, and severely constrained intergroup social relations. The Jim Crow economic system tied black farmers to white landlords in debt peonage that was only one step above slavery. For Mexicans in the 1920s, the new economic system replaced the *patrón-peón* relations in cattle ranching with those more appropriate to tenant farming and agribusiness.[14] These rebuilt caste systems ended brief eras of relatively open opportunity for blacks during Reconstruction and Mexican Americans before the 1920s, ushering in generations of denied rights and repression.

The history of black education in the South is particularly instructive. Immediately after the Civil War, northern Reconstructionists opened schools in the South for freed slave children. The children quickly grabbed the opportunity, so much so that school attendance among the former slaves compared well to school attendance among white children. With the advent of Jim Crow, most of that progress was reversed. Sociologist Stanley Lieberson has shown that, starting about 1880, spending on black schools in the South, where most blacks lived, plummeted from rough parity with spending on white schools to about one-third of the amount spent on white children. School terms for black students became shorter; teacher per student ratios in black schools lower; and black teachers' standards weakened. Outside of a few major cities, high school education was, for all practical purposes, unavailable to black youth in the early years of the twentieth century. In 1911, for example, Atlanta had no high school for black students; in 1930 about one-third of counties in the South had no four-year high schools for blacks. The dismantling of black education was especially devastating in those regions with the greatest concentrations of blacks.[15] Black teenagers today, for the most part, have grandparents who went to those blatantly inferior southern schools.

Starting at the end of Reconstruction, 1877, for African Americans and including Mexican Americans by the 1920s, segregation, sometimes legal and sometimes informal, became the primary means through which access to schooling, housing, and jobs was controlled.[16] (One justification authorities offered for segregation was that mixing of the races would compromise the superior culture and intellect of white Europeans.) American apartheid[17] began with schooling. "Separate but [supposedly] Equal" was the law in many states. No matter where they lived, African American and Mexican American children had to attend separate schools, schools invariably inferior to whites' schools. This profoundly undermined the education as well as the self-esteem of those who were forced to attend. It is not surprising that black and Mexican American children lagged behind whites in school achievement. (Note, however, that in the early part of the twentieth century *northern* blacks did better in school than did most white immigrant groups.)[18] Authorities then used lower academic achievement, a result of educational segregation, to further justify segregation.

Barred by segregation from good education and from well-paying jobs, as well, blacks and Mexican Americans had to live in ghetto housing. The consequences of multiple segregation were profound. It reduced living standards and housing quality and reinforced the segregation of the schools. It concentrated disadvantage into tight quarters. Not the least, segregation told all black and Mexican American children that they were not as good as whites, and this left a psychological scar of inferiority that would have enormously negative effects for decades to come.[19]

Racial stratification remained governed by legal segregation until after World War II when the Civil Rights movement steadily challenged the system's constitutional basis. This challenge culminated in the successful case of *Brown* v. *The Board of Education of Topeka, Kansas*. The Supreme Court reasoned in its decision that "separate but equal" schooling was inherently unequal, that when a society legally separates people by race, it sends a message that they are *not* equal.

After the *Brown* ruling, the tearing down of Jim Crow, and the passing of subsequent civil rights legislation, many Americans thought that both blacks and Mexican Americans would now have the opportunities they had been denied for so long, that a level playing field had finally been established. Progress *was* made. New policies such as equal opportunity enforcement made a difference. In 1940 employed blacks earned 43 percent as much as did whites; by 1980 that proportion had risen to 73 percent. But economic progress leveled off after the mid-1970s.[20] In some areas, including education and occupational advancement, moderate advances have reversed. This new trend troubles Herrnstein and Murray. They point to the growing number of blacks (and others) who are very poor, have been that way for a considerable time, and have little hope of improvement—the so-called underclass. As we noted in earlier chapters, Herrnstein and Murray combine an economic explanation, that low-skill jobs are disappearing, with an explanation based on intelligence, that minorities are insufficiently intelligent to make it in the new marketplace, to explain the group disparities in economic fortunes. (Of course, relatively few blacks and Latinos were succeeding when "suitable" jobs *were* available.)

This is an explanation that ignores history except to claim that history is over. It assumes that three centuries of steady physical, economic, social, and psychological oppression—a record that virtually no one contests— amount to nothing, disappear, evaporate in thirty years. How historically and sociologically naive! Parents' advantages—property, learning, personal contacts, practical crafts, social skills, cultural tools, and so on—are passed on to their children and are also passed from older members of a community to younger ones. Disadvantages are passed on, too. Jim Crow was banished only one generation ago; its legacy will last much longer.

ETHNIC INEQUALITY TODAY

To many, it might appear that the civil rights legislation of the 1950s and 1960s created a level playing field for all, but there are at least two problems with that impression. First, for a century or more, African American and Mexican American families faced severe discrimination in education, housing, jobs, and other economic opportunities. Such disadvantages cumulate and burden future generations. It was naive to think that blacks and Mexican Americans could immediately compete effectively with whites in the labor market. Second, the conditions creating deprivation never fully changed (see box, p. 182). Although de jure segregation has been officially terminated, de facto segregation and discrimination clearly continue.

Critics of government action on racial matters, including the authors of *The Bell Curve*, argue that discrimination against blacks and Latinos *used* to occur but does no longer. If discrimination is gone and these minority groups still lag behind economically, that must be because they lack the ability or will to succeed. But, despite new policies and laws that make discrimination illegal, it persists. News stories of discrimination, such as the refusal of a Denny's restaurant to serve black Secret Service officers, repeatedly point this out. But systematic research points it out, too.

Field "experiments" have demonstrated that there is extensive discrimination in, for instance, the housing market. In the typical study, black and white researchers, posing as homeseekers with identical credentials, approach realtors, agents, lenders, or landlords. At least half of the time

MARTIN LUTHER KING ON THE CONDITIONS FOR EQUALITY

Martin Luther King, Jr., tried to warn both whites and blacks that they should not think that three hundred years of servitude could be overcome with the changing of laws alone. If African Americans and other minorities were to be given a real chance to compete, something other than merely allowing them to compete would be needed. He wrote:

> Whenever this issue of compensatory or preferential treatment for the Negro is raised; some of our friends recoil in horror. The Negro should be granted equality, they agree; but he should ask for nothing more. On the surface, this appears reasonable, but it is not realistic. For it is obvious that if a man is entering the starting line in a race three hundred years after another man, the first would have to perform some impossible feat in order to catch up with his fellow runner.*

* King, *Why We Can't Wait*, p. 134.

discrimination occurs. For example, the black applicants are not shown the properties the whites are; they are told that there are no apartments left to rent while the white applicants are shown those same apartments; and the blacks are "steered" to black neighborhoods.[21] In a society such as ours, where so much depends on where we live—the quality of our schools, police protection, access to jobs, tax assessments, etc.—being turned away from some neighborhoods and being pushed toward others has profound rippling effects.[22]

Other studies show similar discrimination in job hiring. Employers, except when under strong affirmative action pressure, prefer white to black or Latino job applicants three to one, even when qualifications for the job are the same.[23] In one study, businessmen admitted as much. Sociologists Joleen Kirschenman and Kathryn Neckerman interviewed employers in the Chicago area and found many who said that they preferred not to hire black men.[24] Even managers of fast-food places in Harlem, according to another study, prefer to hire nonblacks.[25] In interviews with the *New York Times*, foremen admitted to hiring whites over blacks for construction jobs. One foreman explained, "They [whites] are the people I know best."[26]

Employment discrimination also appears in that realm of supposedly open competition, sports. The days of explicit segregation are gone, but black athletes still have to be just a bit better than white ones to play profes-sionally. There is also some evidence that, because of fan preferences, black players' salaries are lower than they would otherwise be.[27]

And then there are the everyday harassments and slights that identifiable minorities suffer in America: being ignored by storekeepers, watching people cross the street to avoid you, subtle rejection, being questioned by police. Survey research shows that white Americans *are* definitely less prejudiced than they used to be.[28] But blacks do not need to be constantly victimized, as they were under Jim Crow, for discrimination to warp their lives. The occasional encounter with discrimination and a pervasive sense of disparagement, even if now oblique, is enough. Experimental studies show that whites tend, even if unconsciously, to ignore blacks in need or to convey negative impressions of them.[29] (Were only one in eight whites bigoted, that would still leave one hostile white for each black in America. Numbers like these would leave whites with the reasonable impression that bigotry was rare and yet leave blacks with frequent experiences of big-otry—one way that racial groups can experience the same reality so differ-ently.) While each discriminatory incident may be rare and perhaps even trivial, a lifetime of such incidents mounts up to a constant experience of mistreatment.[30]

We might wish that history's heavy weight were the only burden that minorities carried today, but they still confront active discrimination. No-where is this more blatantly visible than in the housing market. The dis-criminatory practices we just described discourage blacks and Latinos from finding housing in white areas. When they have found housing in white areas, it has typically precipitated white flight. Housing segregation means that minority renters and home purchasers pay more than whites for the same housing stock. More critically, it forces many blacks and Mexican Americans to face the same situation that they faced during the era of Jim Crow—segregation.

One of the striking changes in American cities during the twentieth cen-tury has been the *increasing* residential segregation of blacks. The number in figure 8.2 is called the "black residential isolation index." Low numbers mean that blacks tend to live among whites. High numbers mean that blacks tend to live only among other blacks. A perfect 100 would mean that every black lived solely among other blacks. The data cover eighteen large, northern cities. (There are three line segments in the figure because the calculations in different eras had to be based on somewhat different defini-tions of "neighborhood," but the overall trend is clear.) The figure shows that from 1890 to 1970 blacks became increasingly segregated from whites in northern cities.[31] Since 1970, black isolation has leveled off at a high

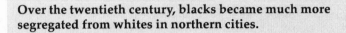

Over the twentieth century, blacks became much more segregated from whites in northern cities.

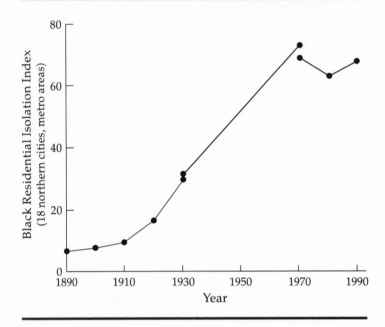

8.2. Index of Black Residential Isolation, Eighteen Northern Cities, 1890–1990 (*Source*: Calculated from Massey and Denton, *American Apartheid*, pp. 24, 28, and 64; and from Harrison and Weinberg, "Racial and Ethnic Segregation")

rate. In the early years, blacks were a small percentage of the population and lived in pockets scattered around the cities. As years passed, more southern blacks moved in and whites fled to other neighborhoods while preventing blacks from following them. The result was the division of northern cities into large, virtually all-black inner neighborhoods and virtually all-white outer neighborhoods. During the same decades that white immigrant groups became *less* segregated from native-born whites and from one another, blacks became *more* segregated.[32] Through the 1980s, poor blacks in particular became increasingly isolated from the wider society as middle-income Americans fled the inner cities. Poor neighborhoods in America have become increasingly all-minority.[33]

The schools that black and Mexican American children attend are segregated again. This time it is de facto segregation attributable to a segregated housing market, but the result is the same. Indeed, de facto segregation may well be even more detrimental psychologically, because African American and Mexican American children can no longer rationalize their separation as the result of legal oppression but see it instead as personal rejection. Although middle-income blacks have been leaving the ghettoes of the poor, they usually end up living in or near low-income black ghettoes, rarely by choice. Their children are often drawn into the world of low-income youth.[34] This experience sends a clear message to all nonwhite youth that their life chances are going to be significantly less than their white counterparts'. What does a good education mean when there is little surety it will bring a good job? What does a good job mean when choice of neighborhoods is restricted?

Even if blacks and Latinos went to schools that were equal in quality to those that whites went to, it would be difficult for teachers to inspire them. Unfortunately, we cannot even test that possibility, because the schools that most of the minority students go to are not equal. Learning in segregated inner-city schools is a difficult task. Many of the physical structures are in poor condition, educational supplies are inadequate, teachers are overwhelmed with disciplinary problems, and, more generally, the climate of pessimism weighs on the talented students as well.[35] Rumors of blocked mobility filter back from the world of work to the schools, fostering discouragement and resentment. It then becomes increasingly difficult for parents, teachers, and school administrators to keep minority students committed and focused on their studies. In this climate, high dropout rates among Latino and African American high school students persist.

Recall anthropologist John Ogbu's distinction between voluntary immigrants and involuntary minorities.[36] Voluntary immigrants, such as Italians, Japanese, and West Indians, chose to come to America for its opportunities. They contrast their conditions here with those in the homeland and feel hopeful. If things do not work out, they can go home, as millions of immigrants have before them. Their optimism in turn sparks commitment to schoolwork. Involuntary minorities, such as blacks and American Indians, and Mexican Americans in a more complex way, were forced to be minorities in the white man's land. They contrast their conditions here with those of their fellow citizens and despair; certainly for American Indians and blacks there is no realistic homeland to compare with or to return to. (Former chairman of the Joint Chiefs of Staff Colin Powell put the distinction this way: "My black ancestors may have been dragged to Jamaica in chains, but they were not dragged to the United States. That is a far different emotional and psychological beginning than that of American blacks,

DELIVERING IN THE BARRIO
AND DANGEROUS MINDS IN THE GHETTO

The popular movie *Stand and Deliver* is instructive about student aspirations in ghetto schools. In this true story, teacher Jaime Escalante takes a group of Mexican American students in one of the poorest schools in Los Angeles and develops their mathematical skills to such a level that they score well on national exams. An inspiring story, *Stand and Deliver* still contains several dispiriting elements. For one, few teachers are as gifted and dedicated as Jaime Escalante was. He put in extra hours in the mornings, evenings, and weekends with his students, overworking himself into a heart attack. Perhaps more important, even after his students had themselves worked so hard, their success was so unexpected that examination regulators charged them with cheating. Mr. Escalante fought for the opportunity to have them take the test again. They scored high again. Yet how often can students' sincere efforts be sustained in the face of such cynicism and suspicion? The stigma that young nonwhites face is a further deterrent to striving in the classroom or on achievement tests.

Dangerous Minds is another truth-based, popular movie about how a dedicated teacher successfully taught minority students. In the story, LouAnne Johnson reached and helped youth in a crime-ridden black ghetto. Ironically, Ms. Johnson, in her own book, did not credit herself for the students' success. She credited a federal grant that paid for smaller classes and for time to provide students with individual instruction.*

* Mosle, "Dissed."

whose ancestors were brought here in chains.")[37] The pessimism of involuntary minorities sparks resistance to schoolwork. These minority youth might even be described as "rationally" pessimistic. Numerous studies show that the economic advantages of staying in school are not nearly as great for blacks and Latinos as for whites.[38]

The oppressive weight of history and the pressure of continuing discrimination together easily explain why African Americans and Mexican Americans remain behind in the American race for success. Racial stratification in the United States cannot be understood by looking at individual traits such as intelligence; ethnic inequality did not emerge from individual competition. Blacks were captured in Africa, Mexicans lost a war; neither

SMARTS ON THE STREET

The debate about race and intelligence has focused on intelligence as expressed in classroom tasks. As we pointed out in earlier chapters, these skills, essentially formal vocabulary and mathematics, form but a small aspect of people's mental abilities. Paper-and-pencil tests capture poorly, if at all, what it means to be "smart" in life outside the classroom—in business, in personal relations, in politics, and so on. Only by observing people in their real lives can one get a full sense of their wider intelligence. Here is where anthropologists and sociologists have a great advantage over psychometricians. Many of the former have studied people in their everyday settings.

In particular, sociologists and anthropologists have studied the kind of people whom Herrnstein and Murray dismiss as constitutionally "dull"— black and brown residents of low-income neighborhoods, many of them high school dropouts, and some even criminals. Although researchers have described bleak lives in these settings, they rarely have described "dull" residents. Of course, the ethnographers have found poor blacks and Latinos who are foolish and shortsighted, but they have also found poor blacks and Latinos who are wise and discerning, just as with any group.

More striking in the research, however, is how quick-wittedly and shrewdly residents of these communities navigate through the perilous waters of ghetto life. Their "street smarts" often entail the same kinds of sophisticated calculation required of professionals and executives. Gang members who operate as black-market entrepreneurs, young men who "hustle" a living, single mothers who balance limited funds and demanding children, working men who juggle multiple low-paying jobs—these are the kinds of people who perform poorly on the paper-and-pencil tests of the classroom but who nevertheless perform shrewdly on the survival tests of the "mean streets."*

* One vast literature that speaks to these points consists of ethnographic studies done in low-income communities, ranging from older ones like Whyte, *Street Corner Society*, to more recent ones such as Stack, *All Our Kin*; Hannerz, *Soulside*; Anderson, *A Place on the Corner*; and Williams and Kornblum, *Growing Up Poor*. Another applicable literature includes studies done specifically of gangs and delinquents, such as Jankowski, *Islands in the Street*; Padilla, *The Gang as an American Enterprise*; Williams, *The Cocaine Kids*; and Wacquant, "Life in the Zone."

group ended up on the bottom because they lost a fair race or scored poorly on a test. The caste system had its origins in colonization—in European Americans' need for cheap labor and their power to get it, to keep it, and to rationalize their control. The role of intelligence has been primarily to justify the existing racial order rather than to create it.

The conditions faced by blacks and Latinos have not changed as much in the last several decades as we might imagine.[39] Most of the young people in these groups face an educational environment and an occupational future that is simply not equal to that of whites. And conditions for the very poor in the black and Latino communities appear to be getting worse.[40]

Given the harsh past and the daunting present we have described, what is perhaps more remarkable than the persisting gap between the academic performance of blacks and whites is that *the gap in test scores is narrowing*. It is a point Herrnstein and Murray admit grudgingly but must admit nevertheless. Over the last twenty years or so, the white advantage over blacks in various standardized tests has narrowed by the equivalent of several IQ points.[41] That alone should cast doubt on the idea that the group differences are inherent and unchangeable. But let us look further.

ETHNICITY, RACE, AND TEST SCORES

The current economic inferiority of ethnic minorities in the United States is, we have shown, sufficiently explained by their centuries-long suppression in a caste system. Purported differences in intelligence are not the cause of their greater poverty, only a post hoc rationalization for it. Nevertheless, we return now to the question of why blacks and Latinos score, on average, below whites on standardized school achievement tests, since that is what much of the controversy is about. (Although often labeled "intelligence" tests, it is more accurate to say that the exams assess how much instruction students have received; see chapters 2 and 3.) Herrnstein and Murray pull up just short of claiming that the group differences in scores arise from group differences in genes, but others have drawn the conclusions *The Bell Curve* authors merely implied.

Many critics have attacked *The Bell Curve* both for its claim that there are racial differences in intelligence and for the implication that those differences are inherently racial. Critics usually engage the debate within the psychometric tradition (see chapter 2), attacking the quality of the tests (see box, p. 189) or the quality of the statistical controls designed to simulate an all-else-being-equal comparison of the groups. For the record, we will mention some of these and other criticisms. But our particular claim is

Cultural Bias in Tests

One common argument made against racial differences in intelligence is that the intelligence tests are "culturally biased." Scoring well depends on knowing words and information more familiar to white, middle-class people than to blacks and Latino Americans. There is some truth to this claim. The problem of cultural content goes back to the early "Alpha" tests used in World War I, which included questions such as: "The Pierce-Arrow car is made in: Buffalo, Detroit, Toledo, or Flint?"* Similarly, a contemporary test for small children requires that they be familiar with a flag pole and that they know the etiquette called for upon having broken an item belonging to another person.** To be in a subordinate minority is, in part, to be culturally distinct and socially isolated from the majority, thereby reducing the chances of answering such questions "correctly."

Herrnstein and Murray respond to these criticisms by echoing Berkeley psychometrician Arthur Jensen's claim that blacks score below whites on "abstract" test items, too. Thus, the gap cannot be due to cultural content, they argue. What is "abstract," however, is not at all clear. Even memorizing and reciting digits in reverse order—one "intelligence test" Herrnstein and Murray stress—requires familiarity and comfort with numbers.*** Also, as conservative economist Thomas Sowell has noted, it was precisely on such sorts of "abstract" items that European immigrants, such as Jews, demonstrated their supposed dim-wittedness in the early twentieth century.****

* Clarence S. Yoakum, *Army Mental Tests* (1919), pp. 260–61, quoted by Lears, *Fables of Abundance*, p. 220.

** Questions quoted in *Science for the People*, "IQ."

*** See Stephen J. Ceci, *On Intelligence*, esp. chap. 9, "How Abstract Is Intelligence?" Herrnstein and Murray also like to cite reaction-time studies that purportedly show black inferiority. Those studies have, however, been persuasively undermined. See Kamin, "Lies, Damned Lies, and Statistics," pp. 87–89; Irvine and Berry, "The Abilities of Mankind," p. 51; and Nisbett, "Race, IQ, and Scientism," p. 44.

**** Sowell, "Ethnicity and IQ."

broader than that: *Scores on achievement tests are the products of inequality, here and elsewhere in the world.*

We return now explicitly to the explanation we outlined earlier and diagrammed in figure 8.1. Ethnic groups in lower caste or status positions tend to score poorly because their position leads to socioeconomic deprivation, group segregation, and a stigmatized identity, each of which undermines performance on psychometric measures of intelligence. As foundation for these claims, we first show that this pattern of ethnic differences in scores is not special to blacks or Latino Americans today; it existed for other groups earlier in American history and exists for other groups around the world.

Many immigrants faced prejudice and discrimination in coming to America (although not slavery or peonage). As early as the colonial era, Benjamin Franklin objected to allowing Germans—"Palatine boors," he called them—into Pennsylvania.[42] Those who came earlier to these shores often considered those who came later as immoral, subversive, and dull. Around 1900 the president of M.I.T., an economist, stated that the new immigrants—at the time they were Italians, Russian Jews, Poles, and other eastern European groups—were inferior to earlier immigrants. Because of cheaper transportation available at the turn of the century, the new arrivals were the least fit of their kind, were failures, and lacked talent.[43]

By the 1920s the nativists could point to scientific evidence for such charges: New immigrants and their children did worse in school and worse on standardized tests than did "old-stock" Americans. As Thomas Sowell, writing in the conservative magazine *American Spectator*, points out, the conclusion that European immigrant groups were of below-average intelligence "was based on hard data, as hard as any data in *The Bell Curve*. These groups repeatedly tested below average on the mental tests of the World War I era, both in the army and in civilian life." They scored especially poorly on the "abstract" test questions.[44] The test scores of Jews were so low that Carl Brigham, an early scholar whose work was often drawn upon in policy debates during the 1920s, wrote that the results "would rather disprove the popular belief that the Jew is highly intelligent." (In Britain, Karl Pearson, a founder of modern statistics, also discovered that Jewish children scored below gentile children on mental ability.) Generally, immigrants from southern and eastern Europe were assessed as significantly inferior to the "Nordic race."[45] Yet a couple of generations later, the test scores of the new groups had risen dramatically, matching or exceeding those of earlier-arriving white Americans. Sowell explains that low scores reflect groups' positions "outside the cultural mainstream of contemporary Western society . . . whatever their race might be." Southern and eastern European immigrant groups arrived as cultural outsiders but became insiders during the course of the twentieth century.

Sowell's argument is not unlike our own. (Sowell's argument and evidence were presumably known to Herrnstein and Murray, because he first published them a quarter-century ago.) Today, blacks and Latinos are still cultural—and economic and social—outsiders in ways far more profound than was true for the Europeans. Today, black test scores are about where those immigrants' scores were in the 1920s—and they are rising.

Table 8.1 lists a sample of studies from around the world that have examined group differences in test scores. We note the limitations of this table: The information is not complete; the tests and procedures varied considerably from study to study, so there is no simple way to compare the size of group differences; in some cases, a few studies have yielded more mixed findings than those shown here, although virtually none found group contrasts opposite to those shown here; and a narrowing of differences appears to be happening in some nations not unlike the narrowing in the United States between blacks and whites. All this said, the table still captures the general pattern.

The table shows, as John Ogbu has argued, that ethnic groups that are inferior in status and caste position score worse on achievement and "intelligence" tests. Ogbu also noted a tendency, in cases where the subordinate group differs racially from the dominant one, for members of the superior group to explain those differences as genetic.[46] Yet *genetic or racial explanations cannot explain the pattern of group differences.*

A reading of the table shows that ethnicity or race, understood biologically, cannot be the cause of the test-score differences. Particularly striking are the substantial gaps in test scores between groups of the same ethnicity or race in countries like Israel, Japan, and South Africa. In Israel, Ashkenazi (Western-origin) Jews have historically scored higher than Mizrachi (Eastern-origin) Jews.[47] Since the founding of Israel in 1948, Eastern-origin Jews have had less wealth, power, and status than Western-origin ones; the latter have often considered and treated the former as culturally "primitive." Although the differences have been narrowing and intermarriage has been growing, Eastern-origin children still score below Western-origin children in many aptitude tests, with differences as large as two-thirds of a standard deviation. (The black-white gap in America is about one standard deviation.) Moreover, researchers continue to find ethnic differences in scores after controlling for ethnic differences in social class.[48]

In Japan, residents of Korean ancestry are so racially indistinguishable from "pure" Japanese that many Korean youth "pass" in school by hiding their ancestry and taking Japanese names; some continue passing afterward by cutting family ties and fabricating new identities. The motivation to pass is strong because the Japanese have historically discriminated against Koreans and consider the Koreans to be a problem group—dull,

TABLE 8.1
Group Differences Around the World

Country	Status or Caste Position		Test Scores, School Success	
	High	Low	High	Low
United States[a]	Whites	Blacks	Whites	Blacks
	Whites	Latinos	Whites	Latinos
	Whites	American Indians	Whites	American Indians
Great Britain[b]	English	Irish, Scottish	English	Irish, Scottish
Northern Ireland[c]	Protestants	Catholics	Protestants	Catholics
Australia[d]	Whites	Aborigines	Whites	Aborigines
New Zealand[e]	Whites	Maoris	Whites	Maoris
South Africa[f]	English	Afrikaaners	English	Afrikaaners
Belgium[g]	French	Flemish	French	Flemish
Israel[h]	Jews	Arabs	Jews	Arabs
	Western Jews	Eastern Jews	Western Jews	Eastern Jews
India[i]	Nontribals	Tribal people	Nontribals	Tribal people
	Brahmin	Harijan	Brahmin	Harijan
	High caste	Low caste	High caste	Low caste
Czechoslovakia[j]	Slovaks	Gypsies	Slovaks	Gypsies
Japan[k]	Non-Burakumin	Burakumin	Non-Burakumin	Burakumin
	Japanese Origin	Korean Origin	Japanese Origin	Korean Origin

a. The white-black and Anglo-Latino differences are reported in *The Bell Curve* and many other places. On American Indians: see, for example, Church, "Academic Achievement."

b. Research by Richard Lynn discussed in Benson, "Ireland's 'Low' IQ."

c. Lynn et al., "Home Background," presents evidence that among young men in Northern Ireland, Protestants scored higher on an intelligence test than did Catholics. (There was no difference among young women.) Calculations based on their table 4 show that the religious difference in IQ among males persists even after controlling for aspects of parental socioeconomic status. A few newspaper stories on *The Bell Curve* reported that, in Belfast, Catholics scored one standard deviation below Protestants in IQ, but we have been unable to find the source for that claim.

d. Klich, "Aboriginal Cognition and Psychological Science"; Clark and Halford, "Does Cognitive Style Account for Cultural Differences?"

e. Ogbu, *Minority Education and Caste*; St. George, "Cognitive Ability Assessment in New Zealand."

f. Verster and Prinsloo, "The Diminishing Test Performance Gap."

g. Raven, "The Raven Progressive Matrices," esp. fig. 2.

h. On Jews versus Arabs: Kugelmass et al., "Patterns of Intellectual Ability"; news item that, in 1992, 26% of Jewish high school students passed matriculation exam versus 15% of Arab students—*Jerusalem Reports*, January 12, 1995; cf. Lieblich et al., "Patterns of Intellectual Ability." On Western versus Eastern Jews: Gross, "Cultural Concomitants of Preschoolers' Preparation for Learning"; Dar and Resh, "Socioeconomic and Ethnic Gaps."

i. For an overview on Indian caste differences, see Das and Khurana, "Caste and Cognitive Processes." Also, on tribal groups, see Gupta and Jahan, "Differences in Cognitive Capacity"; on Brahmins, Shyam, "Variations in Concentration of 'g' Level Abilities"; and on caste: Das, "Level-I Abilities of Socially Disadvantaged Children," ill-mannered, often criminal. Even the word "Korean" is considered a slur. Children of Korean ancestry have historically done relatively worse in school and on aptitude tests.[49]

Another low-status group in Japan also racially indistinguishable from the majority consistently also scores lower than the majority on aptitude tests. These are the Burakumin, descendants of people who formed a sort of "untouchable" caste in feudal Japan. (They, like India's untouchables, dealt with "filth," such as burials and carcasses.) Burakumin who try to "pass" can be unmasked only by careful genealogical analysis, often conducted preparatory to marriage. Still, the majority Japanese consider them to be inferior and often discriminate against them. The Burakumin, like the Koreans in Japan, are prone to economic failure and social problems. And their children have systematically done worse on aptitude tests than non-Burakumin children in the same schools.[50]

In South Africa during the 1950s, children of English origin scored higher on aptitude and intelligence tests than did Afrikaaner (Dutch-origin) children. The gaps between these two northern European groups ran as wide as a half to a full standard deviation. In the 1960s the English-Afrikaaner differences narrowed, and by the 1970s they seemed to have disappeared. The convergence of Afrikaaner and English scores coincides with the rise of Afrikaaners to power in South Africa after generations of subordination to the English (and before conceding power to the native black Africans in the 1990s).[51]

In all three of these cases, evidence suggests that the test-score gap has narrowed over recent generations as subordinated groups began to move slowly toward political and social parity. A similar trend is noticeable over the twentieth century for African Americans, despite continuing economic and geographical isolation.

Still, the basic differences in the table persist. These results, just like the inferior test scores of eastern and southern European immigrants to the United States seventy-five years ago, cannot be reasonably explained by

---

and Das and Padhee, "Level II Abilities of Socially Disadvantaged Children." Some studies of advanced school students do not show caste differences. Rangari, "Caste Affiliation," showed nonsignificant differences in a small sample. Sandhu, "A Study of Caste Differences," also found no differences, but the sample sizes of "backward" and "scheduled" castes in that study were also quite small.

j. Adamovic, "Intellectual Development and Level of Knowledge in Gypsy Pupils."

k. On Burakumin: Shimahara, "Social Mobility and Education"; on Koreans: Lee, "Koreans in Japan and the United States"; DeVos and Wetherall, *Japan's Minorities*.

inherent genetic differences. The most logical explanation is this: Where ethnic groups exist in castelike or near-caste relationships, youths in the subordinate groups do poorly on so-called intelligence tests and similar academic assessments. They do so because of the fewer material and cultural resources their families have, because of segregation, and because they understand the limitations placed on their aspirations in those societies. Black and Latino American youth in the United States are simply like other youth in the world with subordinate caste status.[52] Let us look more closely at this process in the United States.

Socioeconomic Deprivation

Most critics of *The Bell Curve* concede that, on average, black and Latino youth score lower on achievement tests than whites ones do, but they assert that the difference can be entirely accounted for by the differing family backgrounds of the two groups—that is, by the material and cultural deprivation of minority individuals. On average, minority children grow up with lower income, much less wealth, less nutritious diets, unhealthier environments, worse medical care, and so on than do white children. These conditions impair learning. Black and Latino children also more often come from poor home environments in which learning is difficult. Their families value education[53] but cannot create the physical conditions for it and often lack the kind of social skills to support it (for example, knowing how to help with homework or how to find the best teachers). The result is that minority students are generally less interested in academics, have difficulty focusing on their studies, and are more physically active. Such conditions require the ghetto schools to do more than they would have to do in middle-class neighborhoods, and yet the ghetto schools usually have fewer resources with which to do it. If blacks and Latinos were, on average, equal to whites in social conditions, critics of racial explanations say, the performance gap would disappear, showing us that race and ethnicity are irrelevant to intelligence.

The typical way that researchers evaluate this argument is *statistically* to adjust individuals' test scores for differences among them in social and economic circumstances, simulating a situation in which minorities and whites faced equal disadvantages. (See appendix 2 for an introduction to multiple regression analysis.) Herrnstein and Murray (p. 288) perform a version of this procedure when they show, in a graph that was widely disseminated by the media, that at every level of "parental socioeconomic status," from lowest decile to highest decile, blacks scored below whites on the Armed Forces Qualifying Test. From that exercise, they concluded that social background cannot account for the race gap in test scores.

One problem with that work, as we pointed out in chapter 4, is that Herrnstein and Murray's statistical treatment of parental status is faulty and so provides an inadequate test of the argument that differences in deprivation explain differences in scores. Another problem is that there are many more conditions besides parental education, income, and occupation that disadvantage the minority children and that also affect learning. One such condition is wealth. Annual income differences between whites and blacks do not capture the true scale of the differences in wealth. This failure arises, in part, because annual incomes fluctuate widely, because income figures do not capture financial help from kin, and because even small annual gaps in income accumulate over the years. So, for example, the median young, two-earner, black couple brings home 81 percent as much annual income as the median young, two-earner, white couple. But the black couple's net worth is only *18 percent* as great as the white's.[54] Other differences between the races not captured in the typical analysis range from rates of breastfeeding, to exposure to lead poisoning, to parents' clout in the school system.[55]

In a recent study of the young children of the same survey respondents that Herrnstein and Murray analyzed, Jonathan Crane found that the black-white gap in math and reading scores could be totally accounted for by the following differences between black and white children: family income, size of household, proportion of students in the school the mother had attended who were poor, the age the child was weaned, whether the child was read to, and, most important, how much the home was emotionally supportive and cognitively stimulating. Black and white children similar to one another in these conditions performed similarly on the tests. Consequently, Crane concludes that genetics is irrelevant to explaining the test-score gap.[56]

Such studies are controversial, however. The basic charge against them by racial theorists is that simulating equal social circumstances by statistical controls is misleading. These circumstances are *themselves*, psychometricians such as Arthur Jensen contend, the product of low intelligence. For example, parents with low intelligence provide poor home environments for children; those parents are disproportionately black. Statistically eliminating the effect of the home environment masks the fact that poor intelligence breeds poor intelligence genetically; the poor home environment is just a by-product of low parental intelligence, they argue. In turn, researchers who claim that the racial gap can be explained by the individuals' personal environments have rejoinders to this charge.[57] The debate will continue. Certainly, it is fair to conclude that some, perhaps most, of the minority-white difference in scores arises from impairing social conditions, such as the poverty that black and Latino children more often face for longer periods than white children do. But these disadvantages of family class position are not the only ways that being in a lower ethnic caste impairs test performance and may not be sufficient to explain the test-score gap.

Segregation and Isolation

As we have shown, African Americans are severely segregated from white Americans; Latino segregation is less severe but still sizable.[58] Social segregation is also great. Friendships usually do not cross racial lines, and black-white intermarriage is rare. Schools, especially in the larger cities, are also highly segregated.

Residential, social, and school segregation is so profound, especially for blacks, that it often overrides middle-class advantages that some minority children may have. Sociologists have found that the more education or income whites have, the better and safer the neighborhoods in which they live. But this is not so for black Americans; for them, more education or income does not translate as well into better or safer neighborhoods. The major reason is the subtle and not-so-subtle housing discrimination we discussed earlier.[59] Therefore, middle-class blacks often have to live in or near low-income communities, something middle-class whites need not do. Similarly, poor black families are concentrated together much more than poor white families are.[60] This means that poor black children whose parents strive to help them academically must still live and learn with children who lack such social support. Again, equally poor white children are much less likely to have to deal with poor neighbors and disadvantaged fellow students.

Segregation impairs school and test performance in ways that are not revealed by analyses of individual traits. For one, schools in segregated black neighborhoods tend to be poor schools, and differences among schools do affect learning and test scores (see chapter 7). Research shows that minority students, even with the same amount of schooling as comparable white students and even in nominally academic tracks, are less likely than whites to have had classes in advanced mathematics or science.[61] Minority children score low on standardized tests such as the AFQT in part for this reason, because the schools they must attend do not expose them to important curricula. Furthermore, most blacks and Latinos must attend schools with higher classroom size and fewer and older facilities.[62] Many teachers who are creative leave inner-city schools because they tire of the daily battles, leaving the ghetto with an oversupply of teachers who are burnt out or who are inexperienced. Such differences affect students' abilities to score high on standardized tests. In addition, persistent de facto segregation creates conditions that lower youths' self-esteem and ultimately lower their test scores. Blacks and Latinos have to live among increasingly impoverished and marginalized members of their groups; they must cope with the disaffection and disruption of poorer students. Comparable white youth, in contrast, do not face such barriers to learning, because they are not barred from communities of their choice.

And then "concentration effects" accentuate the problems. Much research has shown that people are influenced by the social climate around them, influenced, for example, to vote in ways one would not expect given their individual characteristics. Similarly, research shows that young people are strongly influenced by their peers (no surprise to parents who worry about the crowd with whom their children hang out). In neighborhoods and schools with high concentrations of delinquent youngsters, more youth get into trouble than would be expected given their individual backgrounds.[63] Simply put, even "good" kids can turn "bad" if the setting they are in is heavily comprised of troubled youth and the social climate is negative. Minority children who might be otherwise destined to do well academically do not do well because they are in segregated settings, with concentrated problems, that lead them to learn less.

Together, the deprivation and the segregation of blacks and Latinos go a long way to account for the persistent difference in scores, without needing to appeal to genetics. But we suspect that a residual difference remains. That final bit of difference cannot be erased statistically—nor probably in reality—by raising the economic conditions of individual blacks and Latinos, or perhaps even by moving them into desegregated communities. The residual difference emerges from the fundamental identity of the *group*, African American or Latino American. The profound reality of being black or Latino in America is to be in a lower caste position. That identity creates specific expectations, anxieties, and reactions. These, in turn, handicap youth in test taking and in more important tasks as well.

Stigmatized Identity

Youths who see themselves as fated to a lower caste position come to accept the description of them the majority provides. Or they rebel against it in self-destructive ways. Or they do both. Thus, many nonwhite youth, whatever their ability or learning, score poorly on intelligence and achievement tests.

Nonwhite youth have heard that they are unintelligent and fated to fail; many come to fear that this is true. After all, examples of failure permeate their low-income neighborhoods. The fear affects their confidence and their preparation. When tested, then, they find that the results confirm what others, and sometimes they themselves, believe. Experiments led by psychologist Claude Steele show this process in operation. He and his assistants gave black Stanford University students a test drawn from the Graduate Record Exam. The researchers raised the specter of racial inferiority for a random set of the black students by telling them that the test measured their personal abilities; alternatively, the researchers reminded them of race by asking them to check off their ethnicity on a questionnaire. They told another set of black students, also randomly chosen, that the study was simply psychological research. The first group of students performed notably worse than the second group did. Steele explains that when black students are explicitly confronted with the stereotype of black intellectual inferiority, the resulting anxiety—"Am I going to reveal blacks' inferiority?"—interferes with performance and so they do perform worse. If this happens to black youth who have made it all the way to Stanford, one can imagine that the process is yet stronger among other black youth. (Reinforcing Steele's theory of "stereotype vulnerability" are studies that showed that the same process occurred among women and among white men taking math tests. When told in advance that women usually do worse than men on the test, female undergraduates scored poorly; but when told that women did equally well, the female undergraduates scored as highly as male ones did. Similarly, white men performed worse when they were contrasted to Asians.)[64]

Here is one way *The Bell Curve* and similar books help create that which they claim to explain, the lower performance of minorities—by instilling apprehension. As university teachers of teenagers, we are concerned that the popularization of *The Bell Curve* has demoralized our minority students, reinforced nagging self-doubts, and worsened the problem. It may also falsely inflate the self-images of white students.

Black and Latino youth respond to the caste system with fatalism, with anxiety, and sometimes also with hostility. Their parents' and neighbors' experience suggests that, whatever abstract value they place on education, schooling will probably not pay off for them. If the "program" seems to promise humiliation and dead-end jobs, many decide not to go along with the program. They view cooperation with white institutions, such as schools, as capitulating to the enemy. They develop an oppositional culture, one in which, for example, performing well in school is seen as acting "white" or "doing the Anglo thing." Prestige is found in other realms of life, such as sports. Even good students face peer pressure to defy white standards.[65]

In this climate of collective low self-esteem and resistance, school work suffers. Similarly, in testing situations, students might, in a complete show of disdain, haphazardly answer test questions. (When one of the authors, Martín Sánchez Jankowski, taught junior high school in Detroit, he encountered a consistent test-taking pattern among Gypsy children. They would answer the standardized test questions randomly; some deliberately answered them incorrectly. They said they did not care about school, that the only reason they attended was to avoid having their families get into trouble with the law.) There is evidence of such attitudes toward the AFQT among the minority NLSY respondents who took it.[66]

Korean youth in Japan—but not in America—seem to show similar patterns: low academic expectations and a history of disruptiveness in the schools. Some Korean Japanese children see sports as the only arena for success in school. Their parents are also often skeptical that they can succeed through academics. One father told a visiting anthropologist, "I just hope [my son] is physically tough and strong."[67] British sociologist Paul Willis has described such an oppositional culture among white working-class youth in England. (In England, class differences take on some of the quality of caste differences.) Believing that they are doomed to lousy jobs or no jobs at all and that the society demeans them, they denigrate school-work and honor delinquency.[68] Similar reactions have been observed elsewhere, for example, among Eastern-origin Jews in Israel.[69] The patterns we have described are not unique to American blacks or Latinos, but common to lower-caste groups around the world.

In a caste system where race is a master trait defining people's identities, it should be no surprise that race matters. America has for centuries treated people according to their race and it still does, albeit not as severely now. Young people understand their positions in the racial caste system, irre-

Boys, Girls, and Math: A Parallel Story

The idea that males and females differ in their "natural" talent for mathematics seems intuitively plausible to many Americans. Women are unlikely to take math-heavy subjects in school or to pursue careers in such fields. Not surprisingly, the American media are full of speculations that fundamental differences in brain structure or chemistry explain the gender gap in math.

Recent research by David Baker and Deborah Perkin Jones reveals how wrong that intuition is. Using math tests given in 1982 to 77,000 eighth graders around the world, they found that, on average, boys and girls score about the same. But there is considerable variation among nations. In some countries, boys do better; in others, girls do better; and in some, they do about equally well. Girls tend to perform better relative to boys in those countries where more women go to college and more women hold jobs in modern industries. Also, girls in the 1982 test had improved their performance relative to boys since earlier tests were conducted in 1964. The speed with which a nation's girls closed the gap varied with the degree to which women's participation in their country's workforce had grown. In sum, girls' math performance responded to women's career opportunities in their nations.

These and other results support Baker and Jones's interpretation of the stereotypical gender gap:

> If male students are afforded the possibility of future educational and occupational opportunities as a function of their performance in mathematics, then they may try harder, teachers may encourage them more, and parents and friends may help them see that mathematics is a domain of performance they should take seriously. On the other hand, female students, who are faced with less opportunity, may see mathematics as less important for their future and are told so in a number of ways by teachers, parents, and friends. In short, opportunity structures can shape numerous socialization processes that shape performance.*

Substitute "white" for "male," "minority" for "female," and "academic" for "mathematics" in this quotation. The logic is the same.

* Baker and Jones, "Creating Gender Equality," p. 92. For similar results, see Hanna, "Cross-Cultural Gender Differences in Mathematics Education."

spective of their families' education or wealth. Some adopt the resigned stance expected of them, some rebel against it, all probably worry about it. One result is the same whatever the reaction: poorer than expected performance on tests in school.

What about Asian Americans? They score more highly than whites do on standardized tests. Does that point to a racial explanation? No. First, the data on the Asian advantage in intelligence is mixed; at best, the differences are tiny.[70] Second, the great bulk of Asian American youth today are the children of, or are themselves, "voluntary immigrants," quite different from the experience of the "involuntary minorities," blacks and Latinos. Most have arrived since the 1965 liberalization of the immigration laws, many coming with middle-class backgrounds. Third, early in the century, Asians, like Jews, scored below native whites in the United States on tests; their improvement is a result of the change in their social position.[71]

An additional feature of the Asian case does have wider implications for understanding race and academic performance. The success of Asian and of Asian American children in school can be satisfactorily accounted for by how much more time, attention, ambition, and effort Asian children and their families put into education. Ironically, white Americans' disadvantage relative to Asians seems to rest, in part, on the American idea of "natural" talent. White mothers, children, and teachers are much more likely to attribute success in school to innate intelligence than are Asians; Asians instead typically attribute success much more to hard work. Indeed, John Ogbu reports that the Chinese American high school students he interviewed believe that they are better than white students and so *work harder* to prove that it is so. The American belief in "natural" talent leads to passivity; the Asian belief that talents are learned leads to more hard work and better performance.[72] (It will be interesting to see what happens as Asians assimilate into American culture.)

For African American and Mexican American youth, even equalizing family backgrounds and community settings probably cannot completely close the test-score gap, because the problem is rooted in being in a stigmatized caste. *But the gap can be closed further.* As European immigrant groups were accepted in America, their test scores rose dramatically. In some of the societies listed in table 8.1, recent research seems to show smaller test-score differences than earlier studies did. Those changes—for Eastern-origin Jews in Israel, Maori in New Zealand, Afrikaaners in South Africa, even Burakumin in Japan—appear to match the weakening of ethnic caste and status barriers in those societies. For example, intermar-

A Thought Experiment

One can appreciate the importance of race as a social reality in America with a thought experiment once suggested by an economist. We ask white readers to imagine that a scientist developed a potion that would turn a white person black. Everything else about the person would be exactly the same, except that he or she would now look African, permanently. How much would the experimenter have to pay you in order for *you* to take the potion? The answer might be one estimate of what it costs to be black in America even when everything else about the person is the same. (Those who feel that blacks are advantaged these days should, of course, be willing to pay the experimenter for the privilege of becoming black.)

riage, the greatest breach of the caste lines, has increased in these cases. In the United States, however, the caste lines, especially between black and white, remain firm.

Conclusion

The Bell Curve treatment of racial differences in intelligence is inflammatory and destructive. It is also wrong. Yes, blacks and Latinos consistently score below whites on standardized tests. But notions that this is a "natural" difference, one resulting from genetics, are inadequate. Individual blacks and Latinos confront these tests of school- and school-like knowledge burdened by centuries of disadvantages: family histories rooted in servitude, poverty, and cultural isolation. They also carry heavy disadvantages rooted in conditions today: continuing discrimination, low income, concentration in problem neighborhoods, and inferior schools, to name a few. Like other lower-caste groups around the world, their poorer performance in school and school-like situations can be understood as the result of socioeconomic deprivation, segregation, and a stigmatized lower-caste identity. African Americans and Latino Americans score below whites because to be black or Latino in the United States is to be below whites.

But the gap is closing, by the equivalent of several IQ points a generation, as it closed for other groups that moved from the periphery of American society toward its center. If we choose to exaggerate the remaining ethnic differences, to treat them as natural and inevitable, we will slow down the convergence. If we see, instead, that this inequality, like social inequality generally, is under our control, we can choose to close the gap more quickly. Our fate as a multiracial nation is not in our stars, to paraphrase Shakespeare, nor in our genes, but in our hands.

CHAPTER 9

Confronting Inequality in America:
The Power of Public Investment

W̄E HAVE SHOWN that American inequality cannot be explained in terms of people's "natural" intelligence or other supposedly genetic traits. Understanding inequality requires explaining why individuals end up where they do on the "ladder" of success and explaining why that ladder is built the way it is. For each task, those who argue that social inequality is the result of inequality in natural talent—that people get their "just desserts"—have vastly underestimated the importance of the social environment.

While genetically assisted advantages—height, health, being male, looks—affect how high individuals climb, social environment is more important. One reason environment matters so much is that it determines how individual traits, even genetic ones, translate into material advantages. For example, in most societies, near-sightedness would have been a major handicap; in a society with eyeglasses, it is trivial. In societies with severe restrictions on women, being born female is a heavy burden on individual attainment; in more egalitarian societies, it is less so. Social environment also matters because it directly structures the opportunities individuals have. Family circumstances—number of siblings, parental income, cultural advantages, and so on; the quality and quantity of schooling; neighborhood conditions; job opportunities; and other features of the social context significantly boost or hold back the individual, whatever his or her talent (see, especially, chapter 4).

As for the structure of inequality, individuals' native abilities are largely irrelevant. Investments in improving skills, such as the expansion of higher education in the 1950s and 1960s, and major disinvestments, such as reducing health care for infants, can alter the shape of inequality in a society. But these, too, refer to societal policies and structures, not distributions of "natural" talent. Nations and historical eras differ in the degree of inequality they have because their economic, cultural, and political circumstances differ. For example, the gap between well-off and worse-off Americans' standards of living widened substantially in the last twenty-five years, even while American intelligence stayed constant or increased (see, especially, chapter 5). To understand *systems* of inequality, we have to think beyond individuals to social structure.

Moreover, both operations of the social environment—the way it structures the ladder of inequality and the way it provides advantages for some individuals and disadvantages for others as they clamber up that ladder—are themselves shaped, in part, by political choices (see, especially, chapters 5 through 7). Those choices concern the rules of the marketplace, the way we provide schooling and job opportunities, government interventions, subsidies, and taxes. In these ways, we, as citizens, *decide* the inequality, both of opportunity and of result, that our nation will have.

THE MYTH OF "JUST DESSERTS"

These conclusions are not novel. In fact, many have been commonplace in the social sciences and public discourse for decades. Over twenty years ago, in response to an IQ controversy stirred up in the early 1970s, Mary Jo Bane, a noted scholar who later became an official in the Department of Health and Human Services, and Christopher Jencks, an eminent sociologist and policy analyst, summarized the facts known then in an article about IQ "myths":

> IQ tests measure only one rather limited variety of intelligence, namely the kind that schools (and psychologists [more precisely, psychometricians]) value. Scores on such tests show remarkably little relationship to performance in most adult roles.

> The poor are seldom poor because they have low IQ scores. . . . They are poor because they either cannot work, cannot find adequately paying jobs, or cannot keep such jobs. This has very little to do with their test scores.

> Socioeconomic background has about the same influence as IQ on how much schooling a person gets, on the kind of occupation he enters, and how much money he makes.[1]

Since those words, research has, if anything, indicated that the social environment—encompassing family background, schools, and community—is even more important than Bane and Jencks thought.

Why, after all this well-established scholarship debunking them, do the myths persist? Why do Americans—at least, those "opinion leaders" who debate such issues in magazines and newspapers—seem so receptive to the idea that inequality simply reflects individuals getting their just rewards for their "natural" abilities?

There are several possible answers. A simple one is that it serves many people to believe in the justice of current inequality—especially readers of books like *The Bell Curve* who are (as Herrnstein and Murray repeatedly note) overwhelmingly advantaged members of society. Bane and Jencks suggested that, "like the divine right of kings," the myth that genetic inequality explains economic inequality "help[s] legitimate the status quo."[2] Certainly, it is in the interest—material and ideological—of the "haves" to endorse the theory that inequality is predetermined and unchangeable. But we believe that the bases for belief in the "natural inequality" myth are deeper and more genuine than that.

Receptiveness to messages like that of *The Bell Curve* may be particularly high in eras when inequality widens, becomes blatant, and cries out for explanation. The 1920s, for example, was such an era. The economic boom of the decade benefited the affluent, especially stock speculators, so much that inequality widened considerably. And in that decade eugenics flourished and was applied against "darker" European immigrants. We are now in a similar era. Receptiveness to notions of natural inequality may also be high now because of the widespread impression that efforts to meliorate inequality—antipoverty programs, affirmative action, compensatory education, etc.—have "failed." Whether they really have failed, or were even really tried, is a matter of intense and serious dispute.[3] But it is the opinion of the journalists, politicians, policy intellectuals, and talk-show hosts that seemingly matters here. If social intervention appears to have failed, then opinion leaders are more likely to also believe that inequality is fated and immutable.

Yet another part of the answer is that this social theory is consistent with longstanding American beliefs about inequality.[4] Those beliefs generally hold—although there are, as in any belief system, contradictions and qualifications—that inequality of *result* is perfectly acceptable, so long as it is the outcome of a fair contest in which all contenders have had equality of *opportunity*. What citizens of the new American nation in the nineteenth century resented about the Old World was not so much its inequality of wealth—after all, the leaders of the Revolution were wealthy—as that Old World inequality was "unfair." Both the privileges that decadent aristocrats received by luck of birth and the interference of autocratic governments in the economy robbed hard workers of the "fruits of their labor."[5] In true free-market spirit, Americans wanted—still want—everyone to get out of the way and let the race be run.

If this is how people want their society *to be*, it is a simple step to believe that this is how, for most part, their society *is*. (Most people are not ideologically consistent and can say in one conversation that there is equality of opportunity in America but then say in another conversation something like, "Rich kids get all the breaks." The context of *The Bell Curve* debate and the social position of those who read about it will favor the first response, that America offers equal opportunity. Then, inequality must be the result of natural talent.) To answer Left critics of the American status quo, Herrnstein and Murray and their ideological allies offer a robust statement that American *equality of opportunity* exists and thus American *inequality of result* is fair.

Linked to these beliefs about inequality is the notion that what a fair contest reveals is each person's "natural" gifts. This is not logically necessary; after all, people can win or lose a fair race because of the training their parents gave them. But American ideology treats the "real" individual as the "inner person" standing apart from social context, from family or community. A fair race, then, rewards people for their unique traits, not for what their parents or friends gave them. This truly unique person can only be the "natural" person, the person composed of in-born traits.

We can now see that American beliefs about inequality and talent are rooted yet more deeply in American *individualism*.[6] This individualism, despite its virtues, makes it difficult even to discuss matters of social structure (the "ladder of success" we have referred to) because Americans usually interpret events as the result of individual will. If someone is rich or poor, it is because of who that person is or of what he or she did. Americans tend not to explain outcomes in terms of the circumstances people face.

One effort we have made in this book is to explain how social structure does matter, individual traits notwithstanding. Move a child from a chaotic and impoverished school to an ordered and affluent one, and more often than not that same child will learn more; he or she will become "smarter." Move a job-seeker from a region with a 10 percent unemployment rate to one with a 3 percent rate and more often than not that same job-seeker will land a better-paying position. Move a family headed by working but poor parents from a society with minimal family support to one with family allowances, universal medical care, and other assistance, and more often than not the children in that same family will be healthier, do better in school, and contribute more as adults. In the big picture, one that recognizes the differences among historical eras and societies, the greatest differences are from context to context, rather than person to person.

207

AN AMERICAN STORY: THE SMITHS

These points can be illustrated with the story of three generations of Americans. The names and kinship are fictional. But the stories are true, representative, and instructive.[7]

John Smith, Sr., was born in 1915 in Trenton, New Jersey. His father was a truck driver and his mother had worked as a laundress before his birth. In 1929, at age fourteen, John dropped out of school and started working in a steel mill. The next year, with the onset of the Depression, he was laid off, so he returned to school. John graduated in 1934 and found a job in a foundry—not the kind of job a high school diploma would seem to deserve in those days, but any job in the Depression was a good one. Some in John's cohort were lucky to land jobs in New Deal public works projects. In 1936 John married; in 1938 his oldest son, John, Jr., was born and John, Sr., was laid off from the foundry. Times were no doubt hard and not much help could be expected from John's parents, with their low income and need to hoard for their later years. After twenty-one months of unemployment and as the European war unfolded in late 1939, John found a job in a steel mill. Work was steady through and beyond World War II. (John was too old for military service and was a father.) With growing prosperity, John was able to have two more children after the war and to buy a home. In 1985 John retired. His and his wife's pensions (she had worked on-and-off for state government) and their social security checks provided an annual income of $15,000 in 1990 (about twice the poverty line for a couple).

John, Jr., born in 1938, also grew up in Trenton. Unlike his dad, he continued his education straight through and attended Rutgers University for two years, from 1956 to 1958. His parents paid the tuition of $500 a year and he worked for his living expenses. After his sophomore year John, Jr., took a management-trainee job with a firm in Chicago. The next year he married. In the 1960s the Smiths moved to Richmond, Virginia, where John III ("Johnny") was born in 1966. Having never been unemployed, John, Jr., moved up the ladder through three firms to upper management, making his last move to Atlanta in 1990. In that year, John, Jr., owned a new house in the suburbs and earned $48,000 a year (over $56,000 in 1995 dollars). His wife earned an additional $17,000 as a secretary. Despite widespread corporate downsizing, John, Jr., felt that, at age fifty-two, he had a good future in his new company. And, given that his elderly parents were financially secure, he could look forward to helping out his children.

Johnny, born in 1966, grew up in the Richmond suburbs and attended Duke University, earning a bachelor's degree in political science in 1988. Although his father had paid for $32,000 of his college education, Johnny graduated owing an additional $32,000 in loans. He landed a sales job with a computer firm, trading on his hobby in computers—a hobby that turned out to be more valuable than his college major. In 1991, at age twenty-five—an age when his father already had been married for four years and, even with less education, had been well on his way to corporate success— Johnny was still only a sales rep. He was living in central Charlotte, North Carolina, with a young woman. They had been living together for a few years, were unmarried, childless, and had no expectation of buying a house anytime soon. Johnny earned about $36,000 a year and his companion, working as a substitute teacher, made only $4,000 that year. Johnny was not optimistic about his financial future, putting his prospects of promotion at 50:50.

These three men—grandfather, father, and son—varied hardly at all in their genetic endowments. Even their educational credentials, although different, were roughly representative of their generations. Yet their three careers—humble achievement, solid success, and anxious toehold—differed greatly. They differed because the times, the *social contexts*, in which they lived differed.

Consider how even more dissimilar the stories would be had we focused on three generations of women. As sharply as conditions have changed for men, they have changed much more for women. The typical woman of the 1920s stopped working to marry and raise a large family; the woman of the 1990s more often delays marriage in order to pursue a career. The first might have hoped for a temporary job as a teacher or nurse; the second aspires to a well-paid profession. The woman of seventy years ago could feel financially secure supported by a "breadwinner" husband; the woman of today must consider the real possibility of divorce and single parenting. Once an average woman might expect to spend her widowhood depending on the aid of her children; today's woman can look forward to a financially secure retirement. These have been radical shifts in life, not because of genetic changes, but because of changing contexts.

The differences in contexts are not accidental nor totally a result of forces outside our society. To be sure, such forces—technological change, world competition, war, for example—do partly determine contexts. But much is subject to our control, to our political choices as a national community. We can see illustrations in the three-generation story. Between John, Sr.'s, adulthood and John, Jr.'s, adulthood, all sorts of financial and governmental institutions had been created to reduce the ferocity of the business cycle, so that the younger John did need not to pass through a depression. Also, programs were established that relieved John, Jr., of a weight that had burdened John, Sr.'s, generation, dependent parents. Social security and Medicare directly helped the older generation but also provided material and psychological freedom for the younger generation. Other programs—expansion of higher education, the GI Bill, assistance for homeownership, and so on—aided the ascent of John, Jr.'s, generation. At the same time, contractions of various kinds, some in the world marketplace, others in national priorities, have frustrated John III's generation.

In short, the truly powerful forces in determining why some people get ahead and others do not and why the gap between the two may be wide or narrow are *not* to be found inside individuals but instead outside of them, in the society within which they live.

IMPLICATIONS: WHAT POLICY QUESTIONS SHOULD WE ASK?

Different scientific paradigms make different questions relevant or even comprehensible (see chapter 2). So, too, with theoretical or ideological paradigms. First, consider the implications of the "natural inequality" paradigm, exemplified by *The Bell Curve*. It assumes that the critical trait, intelligence, is *singular*—then people can be ranked on one dimension; that it is *measurable*—then we can efficiently sort people into their appropriate schools and jobs; that it is *genetic*—then one can explain family continuity, legitimate inheritance, and predict talent; and that it is *unchangeable*—then existing inequality must be accepted and efforts to change it are futile. If *any* of these elements fails, so do the policy implications. As many critics have pointed out, for example, even singular, measurable, and genetic traits (such as hair color, myopia, height, and weight) *are* changeable and are changed all the time; only unchangeable traits can justify nonintervention. If one believes that inequality is determined by an innate trait operating in a "free" market, then the following sorts of policy questions make sense:

How can we best assign children to their appropriate slots? What tests should we use? At what age?

How can we reconcile untalented children and young adults to their fates? What can we do to keep their aspirations in line with their limits, so that they are content? Does too much education lead them to frustration?

How do we reorganize democracy to take natural inequality into account? Should voting be restricted to those who test well or succeed in life?

How do we reduce the birth rates of the poor—that is, the "dull"? Is inducement enough, or will it be necessary to compel them?

For the dull who are born, what do we do to sustain them in a humane fashion?

Are moral qualities as well as cognitive ones "natural" in the same way? If so, how do we deal with people born lacking moral instincts? (Several early psychometricians claimed or speculated, for example, that Jews inherited a genetic tendency for dishonesty.)

This agenda is not fanciful. It was explicitly taken up by the psychometricians, educators, and statisticians whose work forms the foundation of *The Bell Curve*. For example, Harvard psychologist William McDougall in 1925 proposed a stratified system of citizenship: Intelligent people would be in class A and have full rights; people who tested poorly or were poorly educated would be in class C and have no vote; and class B would be largely a temporary status for children before they are selected into class A or C.[8] Colgate University president George Barton wrote in 1922 that he hoped intelligence testing would produce a "caste system as rigid as that of India," but one based on the "rational" system of psychology.[9]

Herrnstein and Murray's ideas are distant from these draconian extensions of the "natural inequality" perspective, but they are rooted in the same paradigm. Parts of a similar agenda are hinted at in *The Bell Curve*—for instance, the concern with differential breeding, with helping everyone find the place in society "appropriate" to his or her intelligence, with reconciling people to their limitations,[10] and the suggestion of a guaranteed annual income for the "dull" losers. This agenda need not be undemocratic, racist, or inhumane. But it follows logically from the "natural inequality" paradigm.

The "natural inequality" paradigm, whatever its political implications, does not stand up to the evidence, as this book has demonstrated. Turning to our "social construction" paradigm, which does fit the evidence, a different set of policy questions arise:

How much equality of opportunity do we want? We, as a nation, have set up and can revise the rules for the "race" to success. To what extent do we want those rules to provide every child with an equal start, so that where they end up reflects only their varied talents and not the advantages or disadvantages of their social backgrounds?

The hasty answer might be "as much as possible," but stop to consider what that would really mean: In a society with full equality of opportunity, each child would have the same material advantages, the same challenging school curricula, the same quality neighborhoods, and so forth. That goal directly contradicts another important American value: that parents work hard, scrimp, and save precisely in order to give their children an *advantage*—to provide them with the nicest homes to live in, the best schools to attend, the most supportive neighborhoods, extras like summer camp, and so on. This value is one reason Americans, even those with low incomes, oppose the idea of big inheritance taxes. One of the just rewards of success is to give your children a head start; conversely, one of the costs of failure is being unable to do so.

So the question is a real one: What trade-off do we want between this family value and the value we place on equality of opportunity?

How much equality of result do we want? Once the race is run, what should the rewards be for the winners and losers? As we have documented, the United States is extreme among the industrial nations in its degree of inequality once everything, from market earnings to government programs, is counted. Is that acceptable? Many believe it is, as long as a "safety net" exists, because this level of inequality is equitable and because perhaps it stimulates greater productive energy. Others find American inequality of result intolerably high and unnecessary. Certainly, evidence shows that communities and societies with high degrees of inequality tend to be troubled and torn ones.[11] Critics of inequality also argue that it stunts ability and depresses initiative, thereby reducing productivity. Some in this camp argue that more equality would actually increase the whole "pie" for everyone by bonding people at the bottom into the wider society, reducing ill health and destructive behavior and widening the pool of talent. There is evidence that equality may well spur economic growth (see chapter 5). Realistically, more equality of result would require rewriting the rules of competition and restructuring the system of rewards. Do we want to do that?—and here "we" includes the authors and most of the readers of this book, we who come from the comfortable classes in America.

These are the most general questions that derive from the paradigm presented in this book (they are questions that make no sense in the older paradigm). But other, more specific issues also come onto the agenda. For example:

Which talents should we nurture and how much should we spend doing it? Human skills, whatever genetic component they have, are *developed*. Even children born with serious neurological impediments can today, with timely help, develop into successful adults. So can children born with serious social impediments. It is a matter of identifying the skills we wish to invest in—analytical logic, creativity, empathy, leadership, and so on—and deciding how and how much to invest.

History shows us that public investment, far from undercutting initiative, has unleashed and stimulated American energies. Investments in infrastructure—roads, technology, agricultural development, and the like—undergirded much of the American economic boom. Investments in people—quality schools, public health, higher education, enriched jobs, and the like—have not only enriched us materially but also expanded human horizons. These joint endeavors of the American community have not sapped individual vitality but sparked and sustained it.

How shall we regulate the market? The market is "socially embedded."[12] From standardized weights and measures, to the limited corporation, to work regulations, to tax deductions, governmentally established rules shape the market. So, which ones do we change *if* we want to alter the inequality that arises in the market? Some answer "none" and contend (with little evidence) that certain rules—minimum wages, health and safety codes, required benefits, and so on—already equalize too much and interfere with efficient production. Others answer "a lot." The latter might urge, for example, increasing the bargaining power of labor in setting wages, or raising the minimum wage, or limiting executive income, or making taxes more progressive. Whatever we decide will shape the market, which in turn will shape inequality. Expanding higher education, for example, would reduce earnings inequality; capping the home mortgage deduction would reduce wealth inequality. How do we want to structure the market?

In sum, the paradigm we have presented, one that recognizes that inequality is not "natural" but is socially constructed from our "historical acts," leads to a critical agenda of policy issues for Americans. Because this paradigm is factually accurate, that agenda is a real and urgent one. Whichever way readers may answer the questions on this agenda, they must recognize that Americans *are designing* inequality; even to avoid a decision is to decide for the status quo. Either way, Americans must accept responsibility for the design of inequality instead of blaming nature.

EQUALITY AND OPPORTUNITY

When posed as alternatives, Americans endorse equality of opportunity and reject equality of result.[13] But there is a third option, which might be called full opportunity, that more closely captures the American moral ideal. Jennifer Hochschild of Princeton University has identified four complementary tenets of the American Dream: Everyone can participate; everyone has a reasonable chance for success if he or she plays by the rules;

success and failure is under one's control; and success implies virtue. Inequality on the scale we have today robs some people of the opportunity truly to participate, denies many a reasonable chance to succeed, and handicaps people according to accidents of birth, so that success is not so clearly a badge of virtue. Greater opportunity would bring this country closer to fulfilling the American dream. Securing this kind of opportunity depends, in turn, on our social choices.

Policies that simply promote equal opportunity may not be sufficient to provide full opportunity. One can imagine a science fiction society of impoverished people where a few are chosen by lottery to live in luxury. This equality of opportunity neither rewards nor encourages development of talent. The kind of opportunity Americans want for themselves and their children would require not only equalizing access to resources such as schools, but also public investments to make sure that the schools were excellent ones.

We certainly do not have a society of equal opportunity. The evidence shows how inequalities in family background (especially parents' income) and of the broader social context (including the quality of schools)—despite the difficulty of measuring such contextual factors fully and accurately—shape outcomes, so opportunities are not equally distributed. But research also shows that we could equalize opportunity more, as America did in the 1960s when we expanded higher education. In a stagnant economy, however, or in an economy with growing inequalities in outcomes, simple equality of opportunity may mean insecurity or decline for the vast majority. In such a case, the difference between an unfair race in which the children of the privileged are advantaged versus a fair one may not make that much of a difference if neither the children of the privileged nor of the unprivileged can find secure and rewarding positions.

Policies that would promote full opportunity require extensive public investment. First and most clearly, a society that promoted opportunity would invest in people—in health, primary and higher education, job training, school lunches—whatever would help people develop their talents most fully, so that more, not fewer, people could participate in a modern economy. Second, a commitment to full opportunity would mean investing in job development, so that there would be rewards for people who improved their skills. It would mean extending, not contracting, the kinds of programs that built the broad American middle class—such as low-interest home loans for young families and low-tuition public universities for people of all social backgrounds to go to college. And it would almost certainly mean a different way of providing health care, in a system that encouraged employers to hire more, rather than fewer, workers, and that increased the chances that both rich and poor children had a good start in life. Third, and most central to our argument here, more opportunity necessarily requires a society more equal than the one we have now. If we have a social order in which many people cannot make an adequate living even when working hard or cannot find jobs that provide security or cannot find work at all, then they have little incentive to develop their abilities and no reasonable reward for their efforts. Such a social order inevitably decreases equality of opportunity, too, since it strains marriages, burdens children as they start off in life, and disrupts the communities upon which children depend. So, greater inequality of outcomes necessarily decreases both opportunity and equality of opportunity.

Americans resist what is called "equality of result" because the idea of giving everyone equal benefits no matter what their contributions violates our sense of fairness. But many egalitarian policies stimulate and reward energy and initiative. If American law encouraged higher rates of unionization, more jobs would pay a decent wage. If American government once again regulated corporate wheeling and dealing, more jobs might be preserved. If we provided more social infrastructure, such as affordable child care and longer school years, we would support work and raise childrens' skills at the same time. Even policies of direct redistribution, like health care for poor children or income supplements for poor working families, give children from poor backgrounds a more equal chance in life.

"Equality of result," in an absolute sense, is thus something of a straw man. No one is advocating, nor could one realistically imagine, a society in which people benefit equally regardless of their contributions. But other thriving, wealthy, modern societies stimulate initiative and reward ability with systems of inequality more generous than our own. The real concerns ought to be how much inequality is reasonable and what our social policies do to exacerbate that inequality. At what point does inequality become a powerful drag on our people, giving the poor no real hope of improving their lot, keeping the middle striving ever more frenziedly just to stay in place, and rewarding the rich ever more lavishly, sometimes just for holding assets favored by the tax code?

Naysayers will warn that such moves would undermine our nation's economic health; that inequality is the price we pay for our wealth. Such arguments are in tune with the claims that inequality is "natural." As we pointed out in chapter 1, they are arguments Americans heard more and more as the antigovernment economic policies of the 1980s failed to help the middle class. They were meant to assuage listeners' anxieties about

215

growing inequality. But these warnings are false. We need not tolerate gross inequality in order to have growth. The contrary seems true; such inequality probably slows our growth (see chapter 5).

We have emphasized repeatedly that Americans who participate in our national politics design the inequality with which we live. For example, we determine inequality in the ways we distribute tax burdens, subsidize homeownership, regulate business and labor unions, and finance health care. But the inequality we have does not just affect the overall well-being of our citizens. It also shapes how much *opportunity* our society provides, by whether it encourages individuals to develop their abilities and whether it rewards them fairly for their efforts.

INVOCATION

In closing, we note that any debate over inequality can rest only in part on the weight of the social science evidence. It must also rest on our moral commitments. The explicit moral commitments Americans have made for over two hundred years—despite our frequent failures to live up to them—include a political dedication to equality. Franklin Delano Roosevelt said in 1937 that the "test of our progress is not whether we add more to the abundance of those who have much; it is whether we provide enough for those who have too little." And our commitments include the biblical injunctions to charity. God tells the Israelites, "If there be among you a needy man, one of thy brethren, within any of thy gates . . . thou shalt not harden thy heart, nor shut thy hand from thy needy brother" (Deut. 15:7). And Jesus spoke: "For I was hungry and you gave me food, I was thirsty and you gave me drink, I was a stranger and you welcomed me. . . . Truly, I say to you, as you did to one of the least of these my brethren, you did it to me" (Matt. 25:35, 40).

Summary of *The Bell Curve*

H ERRNSTEIN and Murray begin, in the preface, by raising a concern that Americans are becoming increasingly and more widely divided between a highly educated and well-paid elite at one extreme and an impoverished and problem-beset underclass at the other. How can we understand that?

In the introduction, Herrnstein and Murray define "intelligence" *conceptually* as "a general capacity for inferring and applying relationships drawn from experience" and "a person's capacity for complex mental work" (p. 4), but *operationally* as a person's score on a statistically determined set of test questions, the famed "IQ" test and its surrogates. They defend IQ testing against its many critics. Most scholars, they claim, now agree that IQ tests accurately measure variations in people's "cognitive ability"— colloquially how "smart" they are; that people's IQ scores are stable over their lifetimes; that these tests are *not* biased culturally; and that IQ is 40 to 80 percent "heritable"—that is, that 40 to 80 percent of the variation among people in IQ is due to variation among them in their genes.

Part I outlines the threat of an emerging "cognitive elite." American society today with increasing efficiency draws high-IQ people into its top economic echelons. Since World War II, colleges have accepted and graduated vastly more young Americans and have done so more on the basis of achievement test scores than of family connections. The elite colleges, in particular, have increasingly "creamed" the smartest students. Similarly, occupations that recruit high-IQ people have expanded and perhaps (the case here is more speculative) become more selective of high-IQ candidates. On the job, workers' IQ scores predict how well they perform and do so better than do seemingly job-specific tests or workers' other characteristics. The market, of course, rewards this superior performance by the brightest with the highest incomes. The result is "something new under the sun" (p. 92): stratification of society by cognitive ability, and in particular, the formation of a high-IQ elite. The stratification process is accelerating because the emerging economic system rewards intelligence above all.

The economic separation of the smartest from the rest of us (the authors constantly reassure *readers* that they are among the smartest, not among the rest) is augmented by their increasing physical separation. More and more, the smartest work and live apart from the rest, and smart men and smart women are increasingly marrying one another, which—given the assumption that IQ is largely genetic—makes for even more distinctively smart babies. In the end, we are seeing the formation of a smart, rich, culturally distinct elite "taking on some characteristics of a caste" (p. 113).

Part II of *The Bell Curve* lays the foundation for most of its substantive arguments. In over 230 pages of text, notes, and appendices, the authors try to make the statistical case that individuals' intelligence determines how they live. In particular, people who score low on IQ tests are especially likely to become problematic citizens. By looking at IQ's effects, until now "barely . . . considered" by social scientists, they will be "clearing away some of the mystery that has surrounded the nation's most serious problems" (p. 118). To make their case, Herrnstein and Murray introduce a massive survey and a statistical tool.

The survey Herrnstein and Murray analyzed is the National Longitudinal Survey of Youth (NLSY). Researchers began in 1979 with a sample of over 12,500 youths aged fourteen to twenty-two and have followed them since, asking them a wide variety of questions. In 1980 most of the sample took a battery of tests, including the Armed Forces Qualifying Test (AFQT). Herrnstein and Murray use the AFQT as their IQ test. *The Bell Curve* includes analyses of the NLSY respondents through the 1990 interviews. (In this part of the book, the authors look only at *white, non-Latino* respondents.)

The statistical tool is "logistic multivariate regression analysis." (For a quick introduction to regression analysis, see appendix 2.) Herrnstein and Murray use it for estimating how much the variation in white NLSY respondents' IQs (as measured by their AFQT scores) "explains" variations among them in certain outcomes. How much, they ask, does AFQT score in 1980 explain the chances that an NLSY respondent in the next decade became poor, a high school dropout, unemployed, unmarried, an unwed mother, on welfare, a neglectful mother, or a criminal? To underline the importance of the AFQT score, they regularly compare how much variation in an outcome it explains with how much variation is explained by the socioeconomic status of the respondents' parents. Crudely, parents' class represents "nurture" and the AFQT score represents "nature." (Where they deem it appropriate to testing the effects of the AFQT score, they add other variables to the analysis.)

From chapter 5 through chapter 12, Herrnstein and Murray present the results of statistical analyses claiming that white NLSY respondents' AFQT scores better identify which of them ended up in problematic situations than does their parents' socioeconomic status. In particular, it is the respondents at the extreme low end of the AFQT bell curve who are at great risk. For example, 30 percent of the whites whom Herrnstein and Murray label "very dull"—the bottom 5 percent on the AFQT—were poor in 1989. By contrast, 24 percent of the lowest 5 percent on the scale measuring parents' class position were poor. Using logistic multiple regression analysis (see appendix 2 for a review of regression analysis) to look at the AFQT and parents' class at the same time suggests that the risk of being poor in 1989 was about 25 percent for the "very dull" who had parents of average class position and about 10 percent for children of the very lowest class position who had average AFQT scores. Herrnstein and Murray's conclusion: the AFQT better accounts for adult poverty than does parents' social position—and similarly with the other problem statuses, such as having a work disability, being divorced, and being in jail.

Herrnstein and Murray wrap this section of *The Bell Curve* by giving each respondent to the NLSY a "Middle Class Values" score. A man scored as "middle class" if he had graduated from high school, had been in the labor force throughout 1989, was never interviewed in jail, and was still married to his first wife. ("Never-married people who met all the other conditions except the marital one were excluded from the analysis" [p. 263]. In other words, unmarried people with a problem were kept in the analysis; those with no problems were ignored.) A woman needed to have graduated from high school, have never had a baby out of wedlock, have never been interviewed in jail, and be still married to her first husband. The AFQT, they argue, most strongly explains the likelihood of being middle class by this measure. (Here, as in some other places, the data do not clearly support the claim. Herrnstein and Murray say that even when one looks separately just at respondents who had graduated high school, "a significant independent role for IQ remains" in accounting for variations in the probability of "middle classness" [p. 265]. In fact, the multiple regression table for the 1,162 high school graduates [reprinted on p. 622] shows that the AFQT is *not* significant by the common convention of the discipline [p < .05], while the effect of parental socioeconomic position *is*.) And, so they conclude, "a smarter population is more likely to be, and more capable of being made into, a civil [i.e., educated, working, law-abiding, non-unwed mother, non-divorced] citizenry" (p. 266).

In part III of *The Bell Curve*, Herrnstein and Murray turn to the topic that garnered their book by far the greatest publicity, racial differences in IQ scores. It is not clear why they do so—other than "to try to induce clarity in ways of thinking about ethnic differences on measures of cognitive ability" (p. 270)—since whether or not ethnic groups vary in average IQ has no relevance to their argument about an emerging cognitive elite and a cognitive underclass. (Later, in the section entitled "How Ethnic Differences Fit into the Story," they write, "Given cognitive differences among ethnic and racial groups, the cognitive elite cannot represent all groups equally. . . . A substantial difference in cognitive ability . . . [plays] out in public and private life" [p. 315]. Presumably, they mean that no affirmative action program could equalize representation in the cognitive elite. Logically, still, nothing would be lost from their analysis or their policy prescriptions if this part of the book were absent—as they later admit. See below.)

The explosive chapter 13 reviews the published literature on group differences in IQ scores. That literature, the authors claim, shows that Asians score slightly higher than whites and that blacks consistently score about fifteen points below whites. The authors discuss charges that these findings reflect biases in the tests, and they conclude that the technical literature (largely discussed in an appendix) refutes all critiques. Taking the IQ differences among races as real, the authors turn to explaining them. They report that, in the NLSY, adjusting statistically for their measure of parental social class cuts the black/white gap by over one-third, but they dismiss this exercise because, they argue, the parents' class itself was determined by their IQs, not by their social environment. The black/white gap in the AFQT score persists when one compares blacks and whites from similar social class backgrounds. Although Herrnstein and Murray concede that the black/white gap has shrunk in recent years, they are pessimistic about any future closing of the gap.

In the latter part of chapter 13, the authors present some largely technical reasons for thinking that the IQ differences are genetic in origin, drawing mostly on the work of Arthur Jensen. Blacks and whites, in particular, differ most on the central dimension of IQ scores, what has been labeled "g," a dimension that the literature they cite also claims is mostly inherited. In the end, Herrnstein and Murray argue simply that, at the least, there is *some* genetic component to group differences. They say they are "resolutely agnostic" on the precise mix of environment and genes (p. 311) and, "most important, *it matters little whether the genes are involved at all*" (p. 312, emphasis added). It matters little, because "realized intelligence, no matter whether realized through genes or the environment, is not very malleable" (p. 314); "for practical purposes, environments are heritable, too" (p. 314); and even "knowing that the differences are 100 percent environmental in origin would not suggest a single program or policy that is not already being tried" (p. 315).

In chapter 14, Herrnstein and Murray return to the NLSY and ask the following question: If blacks, Latinos, and whites were hypothetically equal in IQ, how would they differ in outcomes such as college degree, good jobs, unemployment, and unwed motherhood? By adding the AFQT scores to the variables "controlled" in multiple regression equations, Herrnstein and Murray try to show that blacks and whites with similar AFQT scores earn the same income. Put another way, individual AFQT scores "explain" the black/white gap in earnings. "[I]t looks as if the job market rewards blacks and whites of equivalent cognitive ability nearly equally" (p. 325). Also, blacks appear to "overachieve," given their AFQT scores, in graduating from college and in getting good jobs, which Herrnstein and Murray later attribute to affirmative action. Blacks also "overachieve" in social problems. That is, black respondents to the NLSY were likelier than would be expected—given their AFQT scores—to be unmarried, unemployed, unwed mothers, on welfare, and so on. Latinos also tended to have more problems than their AFQT scores would predict, but to a lesser degree. "Racial and ethnic differences in this country are seen in a new light," the authors conclude, "when cognitive ability is added to the picture" (p. 340). That is, "rhetoric about ethnic oppression" must give way to the realization that minorities are getting just what—or even more than—they deserve by virtue of their intelligence.

In chapter 15, the authors of *The Bell Curve* raise the topic of "dysgenesis," a decline in intelligence resulting from duller people having more children than smarter people. They cite research suggesting a decline in national intelligence due to differential fertility. They face a contradiction, however—the "Flynn Effect." This refers to the finding that IQ scores have been rising in the United States (and elsewhere) in recent decades. Herrnstein and Murray step around the Flynn Effect by arguing, first, that it does not reflect real changes in intelligence, but the increasing sophistication of people in taking IQ tests; second, that it may have resulted from the Baby Boom era of the 1950s when even smart people were having many children; and third, that some positive environmental change may have counterbalanced dysgenesis. Using the NLSY data again, they present evidence that the lower-AFQT-scoring women respondents had more children and sooner than did the higher-scoring women; that lower-IQ ethnic groups were reproducing faster than whites. Also, the average IQ of NLSY respondents' *children* (by 1990, many of the young women in the study had become mothers) was lower than that of their mothers. On top of that, the immigrants who are entering the United States today are less smart than the natives (or than the immigrants who came in earlier generations), further lowering the national intelligence. They judge that the national intelligence level is dropping a point or two per generation. Applying the statistics they had earlier calculated from the NLSY, Herrnstein and Murray estimate that a three-point drop in national IQ would have dire consequences, for example, increasing the proportion of people in poverty and of children born out of wedlock by about 10 percent. Indeed, the "social phenomena that have been so worrisome for the past few decades may in some degree already reflect an ongoing dysgenic effect" (p. 365).

Part IV of *The Bell Curve* turns to social policy matters. In chapter 17, Herrnstein and Murray pose the question, "Can people become smarter if they are given the right kind of help?" (p. 390). No, they argue. Pretty much all the ways to use our available resources have been used, and the results generally fall below expectations. They acknowledge some research showing that nutritional supplements for infants and children raise IQ, but they claim that other studies have failed to find effects. Herrnstein and Murray concede that going to school increases intelligence scores (p. 396), but they claim that the sort of variation we see among American schools in, for example, spending or teacher qualifications has small effects on children's abilities. Compensatory education programs have largely failed to close the achievement gap, according to studies the authors cite. (They do recognize, however, a Venezuelan experiment in which Herrnstein participated that raised children's IQs, and they note the partial success of SAT coaching. Both suggest that about forty hours of study can add about three IQ points.) As for Head Start and similar preschool enrichment programs, the authors report that the "consensus is now clear: Cognitive gains vanish before the end of primary school" (p. 745, n. 55). (Again, Herrnstein and Murray concede that there have been a few notable successes, but they raise challenges to those studies.) Adoption studies indicate that moving infants from deprived to advantaged families may increase their IQ scores perhaps a dozen points, but short of that, Herrnstein and Murray conclude, "it is tough to alter the environment for the development of general intellectual ability" (p. 413). It is especially tough given constraints on school budgets. (In the next chapter, the authors add political limits to federal mandates and the resistance of parents and students to harder schoolwork to their list of constraints on the effectiveness of schools.) Maybe something could be done for those at the very bottom in intelligence, but comparatively few children are in that category. (On the other hand, those are the children who presumably will be the problem in the next generation.)

The critical school issue, chapter 18 of *The Bell Curve* argues, is the problem of "leveling." Over the twentieth century, the academic skills of the average American youth have, excepting about a ten-year dip around 1970, improved. However, the verbal SAT scores of college-bound youth have not rebounded since the 1970s (math SAT scores did). Herrnstein and Murray contend that school curricula were "dumbed" down in order to aid the dumbest students; and the dumbing-down worked. A back-to-basics movement recently restored quality to the "hard" disciplines but left humanities and social studies in the hands of levelers and multiculturalists. On top of that, far more money is spent on the disadvantaged than on the gifted, when many of the former will never be able to reach a basic level of education. We should shift funds to supporting the gifted.

In chapters 19 and 20, Herrnstein and Murray turn to the more controversial policy issue of affirmative action in universities and in jobs. The issue belongs in *The Bell Curve*, they say, because proponents of affirmative action assume that group differences can be eliminated. Using college data, they show that black students average lower SAT scores than white students. They argue that the gap shows preferential treatment (but see chapter 2 of this book) and that the size of the gap is not justified by the social goals of affirmative action. Moreover, admitting lower-scoring minority students reinforces stereotypes and undermines their self-esteem. The authors recommend moving back to a more race-blind admissions policy, where group membership would be only a marginal consideration.

Similarly, affirmative action in jobs has been unnecessary and harmful, they claim. Without considering IQ differences by race, it appears that affirmative action policies have had little effect on black employment. Factoring in AFQT score in the NLSY data, however, Herrnstein and Murray argue that blacks have gained better jobs than their IQs merit. Indeed, by the authors' historical reconstruction, blacks had attained job fairness— that is, jobs of a quality proportionate to their IQs—just before the 1964 Civil Rights Act and have gotten more than their fair share since. The cost of "reverse discrimination" is poor performance in jobs and presumably an inefficient workforce. *The Bell Curve*'s policy recommendation is "to get rid of preferential affirmative action and return to the original conception of casting a wider net" (p. 505).

Herrnstein and Murray close *The Bell Curve* with a summary of national trends and advice for national policy. Twentieth-century technology has fostered an affluent aristocracy of the most intelligent, but this "cognitive elite" is increasingly separated from the rest of America: richer, living apart, living differently, thinking differently. They warn, quoting Clinton's secretary of labor, Robert Reich, of the "secession of the successful" (p. 517). "We fear that a new kind of conservatism is becoming the ideology of the affluent . . . along Latin American lines, where to be conservative has often meant doing whatever is necessary to preserve the mansions on the hills from the menace of the slums below" (p. 518). The fate of those below is bleak. Unintelligent mothers doom generations of the poor; the coming job market will provide even fewer jobs for the "dull"; Americans "in the bottom quartile of intelligence . . . will sometime in the not-so-distant future become a net drag . . . [because] for many people, there is nothing they can learn that will repay the cost of teaching" (p. 520); and even a "white underclass" is emerging (p. 520). The prospects are that the American welfare state will expand into a "custodial state" caring for but sternly controlling the growing underclass—"a high-tech and more lavish version of the Indian reservation" (p. 526).

How can we avoid this unpleasant scenario? In the final chapter, entitled "A Place for Everyone," Herrnstein and Murray write that modern efforts to create "equality of outcomes" should be abandoned. "Trying to pretend that inequality [of endowments] does not really exist has led to disaster" (p. 551). Like the Founding Fathers, we should recognize that people differ in natural ability and should again let people find their own "valued places in society" according to their abilities. In a "simpler America," even those of lesser intelligence were valued and nurtured in the local community. That community has in recent decades been "drained" of life by federal centralization. So, more social functions should be restored to the neighborhood. Also, the rules of modern American life should be simplified so they can be more easily understood by the less bright. For example, regulations inhibiting small business should be reduced; the criminal justice system should again clearly link punishment to crime; and family law should again make marriage a prerequisite for parental rights. "The time has come to make simplification a top priority in reforming policy" (p. 546). More concretely, Herrnstein and Murray support cheap and easily available birth control, screening of immigrants by skills, and income supplements for the poor (pp. 546–48). Although "cognitive partitioning . . . cannot be stopped . . . , America can choose to preserve a society in which every citizen has access to the central satisfactions of life" (p. 551).

Statistical Analysis for Chapter 4

THIS STUDY utilized the same dataset analyzed by Herrnstein and Murray, the National Longitudinal Study of Youth (NLSY). Analysis reported below was conducted separately on white and African American samples. The white sample comprises 4,346 individuals who were in the cross-sectional sample and took the Armed Services Vocational Aptitude Battery (ASVAB) test in 1980. The African American sample includes 2,785 individuals who took the ASVAB test in 1980 and were either in the cross-sectional or the supplemental sample. NLSY originally surveyed individuals in 1979, when they were ages 14–22, and scheduled individuals for annual reinterviews through 1991 (the last survey year for which data were analyzed by Herrnstein and Murray). Table A2.1 provides descriptive statistics for all variables used in the analysis separately by race. Table A2.2 provides details of the coding of all variables.

It is worth drawing the reader's attention to several technical issues involved in Herrnstein and Murray's treatment of the data in general and their construction of the "parental SES" composite measure in particular (see chapter 4). The most powerful component of Herrnstein and Murray's index, family income, is deemphasized by the manner in which their index is constructed. Family income is considered only for those individuals who were not independent and reported parental income for either 1978 or 1979. Twenty-one percent of the white sample who did not meet this criteria were, in effect, assigned mean values for the index. By assigning a mean value of 0 to these cases, the overall weight of this z-scored measure when added into the composite measure has a weight less than that of the three other components (mother's education, father's education, and parent's occupation) which in the white sample have fewer missing cases set to the mean. In our analyses we present all components separately with missing data again set at the mean, but also with a dummy variable for this missing data mean-substitution. The extent that the missing data are not average is indicated by the magnitude of the estimates on these dummy variables. In our analyses, we also use a linear family income measure, rather than Herrnstein and Murray's logged measure, since the log transformation does not add to the explanatory power of the models.

Tables A2.3 and A2.4 present results of a logistic regression on the likelihood of poverty in 1990, when individuals are ages 25–33, separately for the white and African American samples, respectively. Models are run nine ways. Model 1 provides results solely on the basis of Herrnstein and Murray's normally distributed "ZZAFQT" test measure. Model 2 presents results from the standard model employed in the analysis presented in *The Bell Curve*—specifically with the following three z-scored variables: ZZAFQT, SES, and age. It is our closest replication of *The Bell Curve* results. Model 3 presents similar results but includes our more complete measures of parental home environment, including number of siblings, farm background, and the four factors that Herrnstein and Murray use to comprise their SES index. To deal with high numbers of missing data on family background and school composition measures, respondents with missing information were assigned mean values and dummy variables were added to account statistically for this adjustment. Missing data on income was assigned two separate dummy variables based on the reason for the incomplete data: Missing Family Income is used for cases where there are no reports of family income; Independent (Missing Income) is used when the respondents reported their own incomes because they were already independent from their families of origin.

Model 4 adds measures of the adolescent social environment, including school composition, region of country, and a dummy variable for those respondents for whom school composition was unreported. This is our most conservative model of the effects of social environment. Model 5 adds a measure for an individual's schooling prior to taking the test—specifically, the years of education completed and whether the student reported being in an academic high school track and thus being exposed to more advanced college preparatory curriculum. Also added in this regression is a separate measure of an individual's schooling after taking the AFQT test. Here, we explicitly take into account formal education, which Herrnstein and Murray excluded from their models. Model 6 includes a range of individual and social characteristics at the time of measuring the labor market outcome—specifically, measures of the unemployment rate of the local labor market and residential location. We contend that this, too, is part of the social environment, but to be conservative, we left contemporaneous environment out of the earlier models. Model 7 adds a dummy variable for gender. Model 8 is the fully specified model which includes current family composition (number of children, marital status and an interaction effect measuring the differing effects of marriage on men and women). Finally, to assess the actual importance of cognitive ability on outcomes in terms of changes to the pseudo R^2 of the model, model 9 provides results similar to model 8 except without Herrnstein and Murray's ZZAFQT measure.

Pseudo R^2 were calculated for the tables presenting results from logistic regressions as McFadden's measure:

$$1 - \ln L(B)/\ln L(a)$$

where $\ln L(B)$ is the log-likelihood of the full model with all variables included and no restrictions, and $\ln L(a)$ is the log-likelihood of the intercept model.

Tables A2.5 and A2.6 present results of the likelihood that a man will be interviewed in jail at some point during the eleven years of the NLSY data collection examined in the study. Models 1 to 3 are identical to those in the previous logistic regressions. Model 4 adds a dummy variable measuring the effect of an individual being in poverty during at least one of the three years from 1978 to 1980. Model 5 adds measures of social context, including school composition, region of country, and a dummy variable for cases where school composition was unreported. Model 6 adds a measure for an individual's schooling prior to taking the test—specifically, the years of education completed and whether the student reported being in an academic high school track. Model 7 provides results similar to Model 6, except without Herrnstein and Murray's ZZAFQT measure.

Tables A2.7 and A2.8 present results of a logistic regression estimating the likelihood that a woman's first child will be born out of wedlock post-1981. Models 1 to 4 are specified identical to those in tables A2.5 and A2.6. Model 5 presents a complete model that also includes measures of social context (school composition and region) and schooling (years of education prior to test and academic track). Model 6 presents the previous model without the ZZAFQT measure to assess the effect of the inclusion of the cognitive performance measure on the pseudo R^2. Models 7 and 8 present results for the earlier models 4 and 5, but with a different sampling specification. In these last two models, the sample (unlike the Herrnstein and Murray one) includes women who did not have any children.

Figures in the text are generated based on a conservative statistical estimate of the social variables. Rather than simply adding the effects of the different components shown in the tables, we constructed a composite weighted measure of the components. This method adjusts for the correlations within the separate components used to measure social background.

A Quick Primer on Regression Analysis

Simple regression analysis is a technique for estimating how much of the variation in one characteristic goes together (co-varies, or correlates) with variation in another. Example: Say we begin with the following question: How do we understand the variation in annual earnings among working people? That is, why do some people make more money than others do? We suspect that people's level of education helps explain some of this variation in earnings. We obtain a list of workers and record the income and years of education for each one, put the data into a computer program, and look at the printout. A simple regression analysis might tell us something like this: More education tends to go along with more earnings, with a "correlation coefficient" of, say, about .40. Indeed, if we assume that education does not just correlate with but actually *determines* workers' earnings, we could say that years of education "explains" 16 percent of the variation in workers' earnings (calculated as .40 × .40). This would leave 84 percent of the variation in workers' earnings unexplained. We might also learn from the printout that, *on the average*, each additional year of education a worker added $1,000 to his or her earnings. So far, so good. But other differences among workers also affect earnings—for example, how many hours they clock in. Moreover, these other differences may also be connected to education and thereby mislead us about the importance of education.

Multivariate regression analysis allows us to sort out that further complexity. We list each worker's earnings, education, and, now, hours worked. The computer returns with something like this: Together, differences in education and in hours explain 25 percent of the variation among workers in their earnings (leaving 75 percent still unexplained). This procedure also allows us to look at the effects of one causal variable while *simulating* a situation in which other causal variables are "held constant" or "controlled"—that is, simulating a situation in which people did not differ on those other causal properties.

Here we learn, hypothetically, that education accounts for only 12 percent of the variation by itself and that each extra year of education produces for the average worker—someone working an average number of hours— $800 more in earnings. (The change from $1,000 to $800 in this example results from the hypothetical "fact" in our illustration that more educated workers also work more hours. In the simple regression analysis, which did not look at hours worked, part of the contribution of hours worked to earnings was statistically attributed to education. The multivariate regression analysis separates out the two presumed "causes.") Conversely, controlling for education this way, we hypothetically find that each hour worked adds $7 to the earnings of the average worker—someone with average education. We can add more and more "explanatory variables" as we try to explain the variation in workers' earnings. Thus, multivariate regression analysis is a technique for estimating the contribution of several presumed "causes" to explaining variation in the distribution of a presumed "effect."

Herrnstein and Murray, in fact, apply a variant of this technique: *logistic* multivariate regression analysis. The effects they are interested in are not continuous, like income, but binary, such as whether or not a person had a job. Logistic techniques adapt multivariate regression analysis to such dichotomous outcomes, but the underlying logic is the same.

A few of the authors of this book are experienced users of this statistical method, but we are also wary of it. To use it well, one must, for example, clearly understand what causes what, measure each "variable" (the property that varies, like education) accurately, apply the appropriate statistical techniques, and so on. If, for example, one measures a variable—say, parents' social class—poorly, that variable will correlate with outcomes poorly and one will be misled into thinking that it is a less important cause than it truly is. The devil is in the details. More generally, even the best-done regression analysis only suggests causal connections. All it really does is tell the researcher about patterns of *correlation*, from which he or she infers causality. It is a weak substitute for direct experimentation on identified causes—as, for example, when a geneticist experiments with removing and inserting a specific gene in a cell. For understanding cause and effect in society, when experimentation is typically impossible, correlational techniques like multiple regression analysis are often all we have as tools. But users must apply it carefully and must extrapolate from it cautiously.

Descriptive Statistics of All Variables by Race

	Whites Only			Blacks Only		
	Mean	S.D	N	Mean	S.D.	N
Test Results						
AFQT	71.688	20.505	4346	46.083	18.469	2785
ZZAFQT (normalized and Z-scored)	0.176	0.897	4346	−0.896	0.794	2785
Individual Characteristics and Social Background						
SES (Z-scored)	0.193	0.884	4341	−0.670	0.944	2763
Family Income (1978–79)[a]	48.309	27.165	3409	24.691	19.486	2337
Parents' SEI (Z-scored)	0.115	0.946	4118	−0.583	0.980	2247
Mother's Education	11.995	2.433	4174	10.769	2.636	2542
Father's Education	12.328	3.270	4067	10.147	3.511	2043
Two-Parent Family	0.902	0.297	4346	0.657	0.475	2785
Pre-1981 Poverty	0.166	0.372	4346	0.573	0.495	2785
Male	0.497	0.500	4346	0.493	0.500	2785
Age	17.557	2.257	4346	17.496	2.195	2785
Siblings (1979)	2.996	1.906	4344	4.751	2.979	2776
Northeast Region	0.223	0.416	4263	0.167	0.373	2747
West Region	0.143	0.350	4263	0.060	0.237	2747
Central Region	0.375	0.484	4263	0.186	0.389	2747
Farm Background	0.058	0.235	4346	0.029	0.167	2785
Independent (Miss. Income)	0.141	0.348	4346	0.097	0.296	2785
Missing Family Income	0.074	0.262	4346	0.064	0.245	2785
Missing Parent's SEI	0.027	0.162	4346	0.050	0.218	2785
Missing Mother's Education	0.040	0.195	4346	0.087	0.282	2785
Missing Father's Education	0.064	0.245	4346	0.266	0.442	2785
School Background						
Years Educ. pre-AFQT	11.806	1.761	4346	11.358	1.696	2785
Years Educ. post-AFQT	1.583	1.968	4313	1.269	1.594	2747
H.S. Academic Track	0.337	0.473	4234	0.276	0.447	2690
School Composition	0.182	0.651	4346	−0.669	1.022	2785
Missing School Reports	0.212	0.409	4346	0.333	0.471	2785
1990 Individual Characteristics and Social Context						
Children in 1990	0.909	1.091	4195	1.086	1.271	2708
Married (1990)	0.616	0.486	3980	0.327	0.469	2550
Married Man (1990)	0.279	0.449	3980	0.155	0.362	2550
Rural (1990)	0.238	0.426	3677	0.186	0.389	2437
Unemployment Rate (1990)	8.374	5.699	3850	7.180	5.196	2469
Central City (1990)	0.096	0.294	3677	0.255	0.436	2437
Individual Outcomes						
No Illegitimate First Birth	0.896	0.306	873	0.360	0.480	486
No Illegitimate First Child	0.947	0.224	1703	0.625	0.484	829
Never Interviewed in Jail	0.980	0.140	2149	0.892	0.310	1329
Not in Poverty (1990)	0.926	0.262	3422	0.739	0.440	1920

a. In thousands, 1990 dollars.

Test Results	
AFQT	Number right on ASVAB sections 2 (arithmetic reasoning), 3 (word knowledge), 4 (paragraph comprehension), and 8 (mathematical knowledge)
ZZAFQT	Normal distribution assigned based on ranking of AFQT # right (see Herrnstein and Murray, pp. 571–73)
Individual Characteristics and Social Background	
SES (Z-scored)	Composite measure based on father's education, mother's education, log of family income and the Duncan SEI score associated with father's occupation (see H&M, pp. 573–75)
Family Income	Average of reported 1978 and 1979 family income in 1990 dollars (see H&M, p. 574)
Parents' SEI (Z-scored)	Duncan SEI in deciles assigned to highest of parents' occupations, -1 assigned to cases with both parents out of labor force (see H&M, p. 574)
Mother's Education	Years of education
Father's Education	Years of education
Male	Dummy variable (coded 1) for men
Two-Parent Family	Dummy variable (coded 1) for individuals living with male and female adults in household at age 14
Poverty Pre-1981	Dummy variable (coded 1) for individuals experiencing a spell of poverty at least once between 1978 and 1980
Age	Age at first interview
Siblings (1979)	Number of siblings in 1979
Northeast Region	Dummy variable (coded 1) for Northeast region with South region omitted, when individual age 14
West Region	Dummy variable (coded 1) for West region with South region omitted, when individual age 14
Central Region	Dummy variable (coded 1) for Central region with South region omitted, when individual age 14
Farm Background	Dummy variable (coded 1) for farm residence in 1979
Independent (Miss. Fam. Income)	No family income reports because individual was independent from parents in 1978 and 1979
Missing Fam. Income	Dummy variable (coded 1) assigned for missing family income data other than due to independence
Missing SEI Reports	Dummy variable (coded 1) assigned for missing parents' SEI data
Missing Mother's Education	Dummy variable (coded 1) assigned for missing data on mother's education
Missing Father's Education	Dummy variable (coded 1) assigned for missing data on father's education
School Background	
Years Educ. pre-AFQT	Years of education completed prior to when individual took AFQT exam
Years Educ. post-AFQT	Years of education completed after individual took AFQT exam
H.S. Academic Track	Dummy variable (coded 1) for individuals in high school academic track
School Composition	Composite measure of school composition based on school administrator reports of the percentage of 10th graders who drop out of high school, the percentage of economically disadvantaged students at the school, and the percentage of nonwhite students at the school. Variables z-scored, summed and divided by three. Mean value of 0 assigned for missing school reports.
Missing School Reports	Dummy variable (coded 1) for missing school reports

(*continued*)

TABLE A2.2 (*cont.*)

1990 Individual Characteristics and Social Context

Children in 1990	Number of children in household in 1990
Married (1990)	Dummy variable (coded 1) for individuals married in 1990
Married Man (1990)	Dummy variable (coded 1) for married men in 1990
Unemployment Rate (1990)	Unemployment rate for local labor market in 1990 based on assignment of mean values to a categorical NLSY variable that identifies six levels of unemployment in three percentage points increments (unemployment rates of greater than 15%, the highest NLSY value were coded 16.5)
Rural (1990)	Dummy variable (coded 1) for not in SMSA 1990, omitted category is in standard metropolitan statistical area but not in central city (or central city not known)
Central City (1990)	Dummy variable (coded 1) for central city residence in 1990, omitted category is in standard metropolitan statistical area but not in central city (or central city not known)

Individual Outcomes

No Illegitimate First Birth	Dummy variable (coded 1) for woman's first birth not occurring out of wedlock after 1981 (sample restriction: women with at least one child; see H&M, p. 604)
No Illegitimate First Child	Dummy variable (coded 1) for woman's first birth not occurring out of wedlock after 1981 (sample includes women who do not have any children)
Never Interviewed in Jail	Dummy variable (coded 1) for men never being interviewed in jail (sample restriction: men only; see H&M, p. 621)
Not in Poverty (1990)	Dummy variable (coded 1) for those individuals not in poverty in 1990 (sample restriction: excluding those in school either 1989–1990; see H&M, pp. 595–97)

TABLE A2.3
Logistic Regression of Likelihood of a Person Being in Poverty in 1990 (Whites Only)

	(A1)	(A2)	(A3)	(A4)	(A5)	(A6)	(A7)	(A8)	(A9)
Intercept	−2.855**	−2.887**	−2.872**	−3.520**	−0.911	−1.165	−0.092	1.473	2.996**
	(0.089)	(0.090)	(0.237)	(0.297)	(0.919)	(0.922)	(0.958)	(1.077)	(0.968)
ZZAFQT[a]	−0.871**	−0.716**	−0.693**	−0.665**	−0.426**	−0.426**	−0.431**	−0.371**	
	(0.077)	(0.086)	(0.088)	(0.090)	(0.105)	(0.105)	(0.108)	(0.115)	
SES[a]		−0.305**							
		(0.080)							
Age[a]		−0.061	−0.095	−0.032	−0.066	−0.062	−0.010	0.009	0.026
		(0.077)	(0.088)	(0.088)	(0.113)	(0.113)	(0.114)	(0.124)	(0.123)
Family Income[a]			−0.404**	−0.392**	−0.377**	−0.369**	−0.373**	−0.338**	−0.352**
			(0.109)	(0.110)	(0.111)	(0.111)	(0.119)	(0.120)	(0.120)
Parents' SEI[a]			−0.051	−0.025	−0.050	−0.039	0.027	−0.004	−0.032
			(0.083)	(0.085)	(0.096)	(0.096)	(0.097)	(0.098)	(0.097)
Mother's Education[a]			0.005	−0.035	0.006	0.004	0.027	0.056	0.033
			(0.088)	(0.090)	(0.092)	(0.092)	(0.093)	(0.101)	(0.100)
Father's Education[a]			−0.057	−0.061	0.005	0.012	0.010	−0.024	−0.049
			(0.095)	(0.096)	(0.101)	(0.101)	(0.103)	(0.111)	(0.111)
Siblings (1979)[a]			0.190**	0.179**	0.149*	0.152*	0.149*	0.047	0.043
			(0.065)	(0.067)	(0.068)	(0.069)	(0.070)	(0.077)	(0.076)
Farm Background			−0.272	−0.348	−0.349	−0.373	−0.242	−0.270	−0.324
			(0.312)	(0.317)	(0.317)	(0.322)	(0.323)	(0.359)	(0.356)
Two-Parent Family			−0.135	−0.037	−0.138	−0.139	−0.125	−0.044	−0.025
			(0.230)	(0.237)	(0.245)	(0.246)	(0.248)	(0.267)	(0.264)
Missing Fam. Income			−0.122	0.017	0.073	0.058	0.007	0.037	0.050
			(0.333)	(0.334)	(0.336)	(0.337)	(0.341)	(0.365)	(0.362)
Independent (Miss. Inc.)			0.298	0.145	0.047	0.008	−0.182	−0.235	−0.283
			(0.225)	(0.230)	(0.234)	(0.234)	(0.238)	(0.263)	(0.262)
Missing Parents' SEI			−0.141	−0.036	−0.180	−0.216	−0.235	0.154	0.232
			(0.431)	(0.430)	(0.331)	(0.332)	(0.337)	(0.473)	(0.473)
Missing Mother's Ed.			−0.023	0.039	−0.042	−0.030	0.024	0.123	0.221
			(0.353)	(0.351)	(0.352)	(0.352)	(0.357)	(0.405)	(0.397)
Missing Father's Ed.			0.314	0.369	0.333	0.354	0.305	0.167	0.150
			(0.267)	(0.267)	(0.267)	(0.268)	(0.271)	(0.309)	(0.306)
School Composition				−0.250**	−0.247**	−0.242**	−0.246**	−0.276**	−0.300**
				(0.067)	(0.067)	(0.068)	(0.069)	(0.075)	(0.075)
Missing School Report				0.465**	0.492**	0.486**	0.413*	0.184	0.188
				(0.165)	(0.168)	(0.169)	(0.171)	(0.191)	(0.191)
West Region				0.909**	0.884**	0.844**	0.875**	0.589*	0.586*
				(0.225)	(0.228)	(0.233)	(0.235)	(0.257)	(0.256)
Northeast Region				0.059	0.114	0.115	0.129	−0.022	−0.046
				(0.276)	(0.278)	(0.285)	(0.287)	(0.309)	(0.307)
Central Region				0.604**	0.608**	0.515*	0.528*	0.313	0.314
				(0.202)	(0.204)	(0.213)	(0.215)	(0.233)	(0.233)

(*continued*)

233

	(A1)	(A2)	(A3)	(A4)	(A5)	(A6)	(A7)	(A8)	(A9)
Years of Ed. pre-AFQT					−0.179*	−0.180*	−0.243**	−0.279**	−0.391**
					(0.074)	(0.074)	(0.077)	(0.083)	(0.076)
H.S. Academic Track					−0.483	−0.471	−0.461	−0.202	−0.320
					(0.256)	(0.257)	(0.257)	(0.272)	(0.268)
Years of Ed. post-AFQT					−0.229**	−0.234**	−0.233**	−0.201**	−0.257**
					(0.068)	(0.068)	(0.068)	(0.071)	(0.069)
Unemployment Rate (1990)						0.029*	0.029*	0.022	0.022
						(0.015)	(0.015)	(0.017)	(0.017)
Central City (1990)						0.492	0.553	0.470	0.415
						(0.267)	(0.269)	(0.284)	(0.283)
Rural (1990)						0.062	0.071	0.081	0.075
						(0.194)	(0.196)	(0.216)	(0.215)
Male							−0.866**	−0.998**	−1.023**
							(0.165)	(0.224)	(0.222)
Children (1990)								0.724**	0.719**
								(0.089)	(0.088)
Married (1990)								−3.074**	−3.112**
								(0.257)	(0.257)
Married Man (1990)								−0.506**	−0.534**
								(0.181)	(0.180)
Pseudo R^2	0.090	0.099	0.116	0.141	0.157	0.162	0.181	0.322	0.315

a. z-scored.   ** p < .01, * p < .05.   N = 3,031.

TABLE A2.4
Logistic Regression of Likelihood of a Person Being in Poverty in 1990
(African Americans Only)

	(B1)	(B2)	(B3)	(B4)	(B5)	(B6)	(B7)	(B8)	(B9)
Intercept	−1.169**	−1.211**	−1.325**	−1.879**	2.210**	1.797**	2.977**	2.529**	3.628**
	(0.062)	(0.064)	(0.131)	(0.159)	(0.680)	(0.696)	(0.736)	(0.829)	(0.777)
ZZAFQT[a]	−0.807**	−0.679**	−0.691**	−0.677**	−0.379**	−0.378**	−0.398**	−0.366**	
	(0.068)	(0.071)	(0.073)	(0.075)	(0.086)	(0.086)	(0.089)	(0.095)	
SES[a]		−0.394**							
		(0.065)							
Age[a]		−0.042	−0.060	−0.043	−0.015	−0.008	0.014	−0.001	0.001
		(0.062)	(0.068)	(0.070)	(0.087)	(0.088)	(0.090)	(0.096)	(0.095)
Family Income[a]			−0.277**	−0.338**	−0.320**	−0.314**	−0.273**	−0.246**	−0.258*
			(0.092)	(0.097)	(0.098)	(0.099)	(0.097)	(0.102)	(0.103)
Parents' SEI[a]			−0.225**	−0.190*	−0.211**	−0.211**	−0.238**	−0.191*	−0.220**
			(0.074)	(0.075)	(0.074)	(0.074)	(0.076)	(0.084)	(0.083)
Mother's Education[a]			−0.115	−0.153*	−0.081	−0.082	−0.042	−0.017	−0.037
			(0.070)	(0.071)	(0.074)	(0.074)	(0.076)	(0.082)	(0.081)
Father's Education[a]			0.015	0.018	0.062	0.070	0.074	0.065	0.058
			(0.074)	(0.075)	(0.078)	(0.078)	(0.080)	(0.086)	(0.085)
Siblings (1979)[a]			−0.056	−0.038	−0.061	−0.066	−0.071	−0.049	−0.027
			(0.063)	(0.064)	(0.066)	(0.066)	(0.068)	(0.072)	(0.072)

	(B1)	(B2)	(B3)	(B4)	(B5)	(B6)	(B7)	(B8)	(B9)
Farm Background			−0.210	0.002	−0.118	−0.199	−0.161	−0.273	−0.232
			(0.354)	(0.355)	(0.363)	(0.367)	(0.382)	(0.428)	(0.422)
Two-Parent Family			−0.072	0.031	−0.046	−0.043	−0.034	0.101	0.124
			(0.139)	(0.142)	(0.150)	(0.151)	(0.154)	(0.159)	(0.158)
Missing Fam. Income			−0.115	−0.085	0.009	−0.043	−0.024	0.023	0.144
			(0.288)	(0.294)	(0.299)	(0.301)	(0.303)	(0.323)	(0.321)
Independent (Miss. Inc.)			0.706**	0.734**	0.608*	0.595*	0.372	0.194	0.196
			(0.225)	(0.232)	(0.239)	(0.241)	(0.247)	(0.264)	(0.262)
Missing Parents' SEI			0.263	0.164	0.235	0.206	0.241	−0.039	−0.015
			(0.260)	(0.268)	(0.184)	(0.185)	(0.189)	(0.301)	(0.302)
Missing Mother's Ed.			0.090	0.115	−0.049	−0.030	−0.003	0.072	0.114
			(0.217)	(0.220)	(0.225)	(0.225)	(0.233)	(0.247)	(0.247)
Missing Father's Ed.			0.199	0.188	0.131	0.144	0.146	0.073	0.095
			(0.152)	(0.155)	(0.158)	(0.159)	(0.164)	(0.174)	(0.172)
School Composition				−0.256**	−0.246**	−0.262**	−0.240**	−0.245**	−0.265**
				(0.069)	(0.070)	(0.072)	(0.073)	(0.078)	(0.078)
Missing School Report				0.569**	0.528**	0.577**	0.522**	0.498**	0.567**
				(0.154)	(0.157)	(0.161)	(0.164)	(0.174)	(0.172)
West Region				0.625*	0.693**	0.718**	0.739**	0.617*	0.655*
				(0.260)	(0.264)	(0.267)	(0.275)	(0.288)	(0.286)
Northeast Region				0.426*	0.347	0.270*	0.347	0.199	0.109
				(0.187)	(0.194)	(0.204)	(0.206)	(0.214)	(0.212)
Central Region				0.818**	0.809**	0.774**	0.812**	0.578**	0.521**
				(0.155)	(0.158)	(0.163)	(0.167)	(0.179)	(0.177)
Years of Ed. pre-AFQT					−0.318**	−0.307**	−0.375**	−0.338**	−0.424**
					(0.056)	(0.057)	(0.059)	(0.063)	(0.059)
H.S. Academic Track					−0.421*	−0.421*	−0.391*	−0.359*	−0.448**
					(0.164)	(0.165)	(0.167)	(0.176)	(0.174)
Years of Ed. post-AFQT					−0.246**	−0.240**	−0.258**	−0.211**	−0.264**
					(0.056)	(0.056)	(0.057)	(0.061)	(0.059)
Unemployment Rate (1990)						0.029*	0.031*	0.031*	0.031*
						(0.012)	(0.012)	(0.013)	(0.013)
Central City (1990)						0.145	0.142	0.043	0.068
						(0.153)	(0.157)	(0.166)	(0.165)
Rural (1990)						0.226	0.214	0.297	0.316
						(0.167)	(0.171)	(0.183)	(0.182)
Male							−0.993**	−0.676**	−0.675**
							(0.134)	(0.170)	(0.169)
Children (1990)								0.424**	0.420**
								(0.061)	(0.061)
Married (1990)								−2.030**	−2.070**
								(0.226)	(0.225)
Married Man (1990)								−0.090	−0.101
								(0.181)	(0.180)
Pseudo R²	0.086	0.105	0.117	0.146	0.176	0.180	0.210	0.289	0.281

a. z-scored.   ** p < .01, * p < .05.   N = 1,726.

Logistic Regression of Likelihood of a Man Being Interviewed in Jail after AFQT
(Whites Only)

	(A1)	(A2)	(A3)	(A4)	(A5)	(A6)	(A7)
Intercept	−4.464**	−4.462**	−5.053**	−5.375**	−6.125**	−0.957	−0.128
	(0.229)	(0.230)	(0.668)	(0.689)	(0.832)	(1.896)	(1.847)
ZZAFQT[a]	−0.940**	−0.950**	−0.825**	−0.737**	−0.663**	−0.396	
	(0.168)	(0.196)	(0.202)	(0.208)	(0.216)	(0.241)	
SES[a]		0.014					
		(0.185)					
Age[a]		0.014	0.001	0.006	0.092	0.404	0.426
		(0.179)	(0.193)	(0.197)	(0.193)	(0.220)	(0.220)
Family Income[a]			−0.338	−0.091	−0.099	−0.054	−0.070
			(0.247)	(0.236)	(0.238)	(0.243)	(0.242)
Parents' SEI[a]			0.221	0.273	0.281	0.313	0.275
			(0.209)	(0.214)	(0.227)	(0.239)	(0.239)
Mother's Education[a]			−0.035	−0.058	−0.152	−0.137	−0.203
			(0.219)	(0.217)	(0.227)	(0.228)	(0.226)
Father's Education[a]			−0.126	−0.125	−0.036	0.049	0.005
			(0.220)	(0.212)	(0.211)	(0.230)	(0.230)
Siblings (1979)[a]			0.172	0.129	0.136	0.100	0.113
			(0.152)	(0.154)	(0.155)	(0.163)	(0.161)
Farm Background			−a−	−a−	−a−	−a−	−a−
Two-Parent Family			0.662	0.751	0.897	0.965	1.060
			(0.655)	(0.668)	(0.702)	(0.727)	(0.731)
Missing Fam. Income			−1.003	−0.958	−1.029	−1.030	−1.037
			(1.033)	(1.033)	(1.047)	(1.060)	(1.062)
Independent (Miss. Inc.)			0.014	−0.132	−0.394	−0.664	−0.795
			(0.667)	(0.675)	(0.709)	(0.730)	(0.726)
Missing Parents' SEI			−a−	−a−	−a−	−a−	−a−
Missing Mother's Ed.			0.364	0.340	0.728	0.343	0.382
			(0.721)	(0.723)	(0.698)	(0.737)	(0.727)
Missing Father's Ed.			0.787	0.841	0.742	0.573	0.636
			(0.578)	(0.587)	(0.576)	(0.590)	(0.586)
Poverty Pre-1981				1.253**	1.260**	1.180**	1.259**
				(0.399)	(0.405)	(0.406)	(0.404)
School Composition					−0.424**	−0.384**	−0.416**
					(0.140)	(0.141)	(0.142)
Missing School Report					1.233**	1.134**	1.106**
					(0.371)	(0.382)	(0.383)
West Region					0.352	0.463	0.477
					(0.545)	(0.555)	(0.552)
Northeast Region					−0.166	−0.058	−0.049
					(0.587)	(0.602)	(0.604)
Central Region					0.322	0.350	0.372
					(0.460)	(0.475)	(0.476)
Years of Ed. pre-AFQT						−0.432**	−0.500**
						(0.155)	(0.149)
H.S. Academic Track						−1.253	−1.426
						(0.781)	(0.770)
Pseudo R^2	0.089	0.089	0.134	0.160	0.212	0.245	0.238

a. z-scored.   −a− insufficient cases to estimate.   ** p < .01, * p < .05.   N = 2,062.

Logistic Regression of Likelihood of a Man Being Interviewed in Jail after AFQT
(African Americans Only)

	(B1)	(B2)	(B3)	(B4)	(B5)	(B6)	(B7)
Intercept	−2.354**	−2.360**	−2.404**	−2.570**	−3.206**	0.206	1.171
	(0.109)	(0.111)	(0.210)	(0.259)	(0.314)	(1.016)	(0.949)
ZZAFQT[a]	−0.721**	−0.691**	−0.664**	−0.651**	−0.612**	−0.416**	
	(0.110)	(0.114)	(0.118)	(0.119)	(0.123)	(0.134)	
SES[a]		−0.102					
		(0.102)					
Age[a]		−0.002	0.050	0.061	0.096	0.260*	0.301*
		(0.102)	(0.107)	(0.108)	(0.110)	(0.121)	(0.119)
Family Income[a]			−0.111	−0.047	−0.069	−0.061	−0.057
			(0.138)	(0.145)	(0.148)	(0.149)	(0.147)
Parents' SEI[a]			−0.075	−0.060	−0.015	−0.044	−0.081
			(0.121)	(0.121)	(0.121)	(0.123)	(0.122)
Mother's Education[a]			0.037	0.044	0.023	0.072	0.051
			(0.117)	(0.117)	(0.118)	(0.121)	(0.120)
Father's Education[a]			0.075	0.083	0.045	0.058	0.042
			(0.123)	(0.123)	(0.126)	(0.126)	(0.125)
Siblings (1979)[a]			0.026	0.007	0.042	0.015	0.036
			(0.100)	(0.102)	(0.104)	(0.105)	(0.104)
Farm Background			−0.316	−0.356	−0.167	−0.321	−0.307
			(0.556)	(0.557)	(0.567)	(0.574)	(0.570)
Two-Parent Family			−0.064	−0.049	0.046	0.059	0.095
			(0.228)	(0.228)	(0.236)	(0.238)	(0.236)
Missing Fam. Income			0.159	0.076	0.048	0.135	0.222
			(0.384)	(0.390)	(0.404)	(0.408)	(0.403)
Independent (Miss. Inc.)			−0.471	−0.475	−0.368	−0.454	−0.503
			(0.562)	(0.562)	(0.566)	(0.571)	(0.570)
Missing Parents' SEI			0.798*	0.782	0.732	0.859*	0.985*
			(0.407)	(0.407)	(0.417)	(0.426)	(0.421)
Missing Mother's Ed.			0.105	0.076	0.070	−0.032	0.056
			(0.308)	(0.309)	(0.313)	(0.316)	(0.313)
Missing Father's Ed.			0.203	0.191	0.279	0.257	0.311
			(0.241)	(0.241)	(0.245)	(0.246)	(0.243)
Poverty Pre-1981				0.293	0.355	0.236	0.279
				(0.258)	(0.262)	(0.265)	(0.262)
School Composition					−0.209	−0.178	−0.209
					(0.114)	(0.113)	(0.112)
Missing School Report					0.762**	0.648*	0.707**
					(0.252)	(0.254)	(0.252)
West Region					1.148**	1.191**	1.191**
					(0.328)	(0.332)	(0.328)
Northeast Region					0.480	0.463	0.388
					(0.270)	(0.272)	(0.269)
Central Region					0.079	0.021	−0.094
					(0.287)	(0.290)	(0.287)
Years of Ed. pre-AFQT						−0.281**	−0.364**
						(0.086)	(0.081)
H.S. Academic Track						−0.712*	−0.858**
						(0.312)	(0.307)
Pseudo R^2	0.059	0.060	0.072	0.074	0.104	0.126	0.115

a. z-scored.   ** p < .01, * p < .05.   N = 1,255.

TABLE A2.7

Logistic Regression of Likelihood of a Woman Having an Illegitimate First Child after AFQT
(Whites Only)

	(A1)	(A2)	(A3)	(A4)	(A5)	(A6)	(A7)[b]	(A8)[b]
Intercept	−2.396**	−2.545**	−1.487**	−1.738**	−1.469	−0.721	−2.765**	−2.713
	(0.137)	(0.152)	(0.333)	(0.349)	(1.652)	(1.595)	(0.335)	(1.579)
ZZAFQT[a]	−0.754**	−0.524**	−0.518**	−0.470**	−0.283		−0.546**	−0.332*
	(0.131)	(0.148)	(0.158)	(0.159)	(0.174)		(0.143)	(0.161)
SES[a]		−0.259						
		(0.135)						
Age[a]		−0.511**	−0.383*	−0.402*	−0.283	−0.253	−0.273*	−0.197
		(0.147)	(0.160)	(0.160)	(0.207)	(0.206)	(0.146)	(0.197)
Family Income[a]			−0.205	−0.058	−0.081	−0.094	−0.165	−0.201
			(0.160)	(0.158)	(0.157)	(0.158)	(0.158)	(0.161)
Parents' SEI[a]			0.160	0.194	0.223	0.198	0.183	0.207
			(0.139)	(0.141)	(0.146)	(0.144)	(0.133)	(0.138)
Mother's Education[a]			−0.126	−0.116	−0.167	−0.198	−0.095	−0.075
			(0.141)	(0.143)	(0.153)	(0.150)	(0.139)	(0.149)
Father's Education[a]			−0.065	−0.053	0.009	−0.046	−0.150	−0.124
			(0.160)	(0.161)	(0.171)	(0.167)	(0.152)	(0.163)
Siblings (1979)[a]			0.235*	0.246*	0.196	0.197	0.219*	0.205
			(0.117)	(0.120)	(0.124)	(0.123)	(0.107)	(0.111)
Farm Background			−0.413	−0.600	−0.831	−0.948	−0.292	−0.518
			(0.643)	(0.654)	(0.665)	(0.663)	(0.625)	(0.633)
Two-Parent Family			−1.059**	−0.984**	−0.840*	−0.782*	−0.738*	−0.623*
			(0.345)	(0.354)	(0.366)	(0.362)	(0.326)	(0.340)
Missing Fam. Income			−0.482	−0.463	−0.477	−0.515	−0.611	−0.539
			(0.545)	(0.546)	(0.555)	(0.554)	(0.533)	(0.539)
Independent (Miss. Inc.)			−1.323	−1.443	−1.553*	−1.623*	−1.317	−1.504*
			(0.759)	(0.762)	(0.774)	(0.776)	(0.752)	(0.767)
Missing Parents' SEI			0.313	0.307	0.716	0.824	0.231	0.611
			(0.679)	(0.679)	(0.685)	(0.680)	(0.640)	(0.656)
Missing Mother's Ed.			0.046	−0.050	−0.017	0.088	0.116	0.215
			(0.570)	(0.599)	(0.608)	(0.599)	(0.547)	(0.557)
Missing Father's Ed.			−0.088	−0.276	−0.057	−0.004	−0.170	−0.083
			(0.502)	(0.528)	(0.528)	(0.519)	(0.477)	(0.485)
Poverty Pre-1981				1.055**	1.081**	1.110**	0.926**	0.969**
				(0.294)	(0.301)	(0.300)	(0.284)	(0.288)
School Composition					−0.135	−0.151		−0.121
					(0.129)	(0.127)		(0.117)
Missing School Report					0.343	0.392		0.429
					(0.301)	(0.298)		(0.278)
West Region					0.878*	0.883*		0.897*
					(0.396)	(0.393)		(0.370)
Northeast Region					0.033	−0.055		−0.256
					(0.475)	(0.471)		(0.447)
Central Region					0.782*	0.770*		0.747*
					(0.363)	(0.362)		(0.335)
Years of Ed. pre-AFQT					−0.062	−0.123		−0.029
					(0.133)	(0.129)		(0.127)
H.S. Academic Track					−1.319**	−1.417**		−1.605**
					(0.496)	(0.492)		(0.486)
Pseudo R^2	0.068	0.099	0.139	0.161	0.199	0.194	0.147	0.197

a. z-scored.   b. Models A7–A8 also include in sample women who did not have children.

** p < .01, * p < .05.   N = 831 (models A1–A6); N = 1,641 (models A7–A8).

TABLE A2.8

Logistic Regression of Likelihood of a Woman Having an Illegitimate First Child after AFQT
(African Americans Only)

	(B1)	(B2)	(B3)	(B4)	(B5)	(B6)	(B7)[b]	(B8)[b]
Intercept	0.611**	0.428**	0.391	0.112	0.490	1.338	−0.980**	−1.506
	(0.099)	(0.110)	(0.248)	(0.280)	(1.596)	(1.487)	(0.205)	(1.150)
ZZAFQT[a]	−0.436**	−0.292**	−0.312**	−0.248*	−0.214		−0.313**	−0.327**
	(0.102)	(0.113)	(0.118)	(0.122)	(0.143)		(0.093)	(0.107)
SES[a]		−0.097						
		(0.110)						
Age[a]		−0.483**	−0.454**	−0.442**	−0.406	−0.362	−0.320**	−0.350**
		(0.125)	(0.136)	(0.136)	(0.211)	(0.207)	(0.098)	(0.155)
Family Income[a]			−0.060	0.027	0.033	0.022	−0.006	−0.017
			(0.120)	(0.128)	(0.135)	(0.133)	(0.103)	(0.105)
Parents' SEI[a]			−0.301*	−0.281*	−0.310*	−0.312*	−0.087	−0.090
			(0.136)	(0.137)	(0.143)	(0.143)	(0.102)	(0.103)
Mother's Education[a]			0.264*	0.271*	0.252	0.242	0.218*	0.213
			(0.125)	(0.125)	(0.132)	(0.132)	(0.099)	(0.101)
Father's Education[a]			0.136	0.121	0.135	0.117	−0.012	0.001
			(0.122)	(0.123)	(0.127)	(0.126)	(0.095)	(0.097)
Siblings (1979)[a]			0.303**	0.270*	0.307*	0.313**	0.263**	0.272**
			(0.116)	(0.117)	(0.121)	(0.121)	(0.086)	(0.087)
Farm Background			−1.627	−1.702	−1.460	−1.548	−2.155*	−2.118*
			(1.152)	(1.164)	(1.183)	(1.183)	(1.052)	(1.057)
Two-Parent Family			0.044	0.059	0.150	0.154	0.061	0.100
			(0.275)	(0.276)	(0.286)	(0.285)	(0.198)	(0.202)
Missing Fam. Income			−0.072	−0.116	−0.104	−0.084	−0.296	−0.331
			(0.516)	(0.525)	(0.539)	(0.539)	(0.378)	(0.382)
Independent (Miss. Inc.)			−0.240	−0.387	−0.391	−0.409	−0.238	−0.190
			(0.540)	(0.542)	(0.561)	(0.564)	(0.448)	(0.452)
Missing Parents' SEI			0.751	0.707	0.600	0.631	0.382	0.320
			(0.478)	(0.485)	(0.503)	(0.504)	(0.328)	(0.333)
Missing Mother's Ed.			0.165	0.274	0.289	0.297	−0.011	−0.015
			(0.505)	(0.505)	(0.522)	(0.516)	(0.345)	(0.348)
Missing Father's Ed.			0.103	0.090	0.041	0.049	0.287	0.246
			(0.288)	(0.289)	(0.300)	(0.298)	(0.207)	(0.210)
Poverty Pre-1981				0.531*	0.525*	0.578*	0.454*	0.485*
				(0.246)	(0.258)	(0.256)	(0.190)	(0.193)
School Composition					0.125	0.105		0.011
					(0.125)	(0.124)		(0.091)
Missing School Report					0.449	0.443		0.184
					(0.273)	(0.273)		(0.192)
West Region					−0.337	−0.373		−0.514
					(0.528)	(0.521)		(0.393)
Northeast Region					0.773*	0.693*		0.418
					(0.339)	(0.334)		(0.227)
Central Region					0.514	0.498		0.424*
					(0.284)	(0.284)		(0.205)
Years of Ed. pre-AFQT					−0.050	−0.121		0.029
					(0.125)	(0.116)		(0.091)
H.S. Academic Track					−0.424	−0.504*		−0.115
					(0.243)	(0.237)		(0.182)
Pseudo R[2]	0.032	0.059	0.097	0.104	0.137	0.133	0.086	0.096

a. z-scored.   b. Models A7–A8 also include in sample women who did not have children.

** $p < .01$, * $p < .05$.   N = 467 (models A1–A6); N = 802 (models A7–A8).

❖ *Notes* ❖

CHAPTER 1

1. Quoted by the Associated Press, "Bishops Say U.S. Neglects Poor," *San Francisco Chronicle*, November 15, 1995, p. A5.

2. A few do. For discussion of the evidence, see chapter 5.

3. The fuller quotation is: "I'm trying to help people on welfare. I'm not hurting them. The government hurt them. The government took away something more important than this money. They [*sic*] took away their initiative, they took away a substantial measure of their freedom, they took away in many cases their morality, their drive, their pride. I want to help them get that back. . . . I want to do it because I love them, because I want them to be Americans. And their children and grandchildren will thank us." Quoted in Frum, "Righter than Newt," p. 84.

4. See, for example, Levy, *Dollars and Dreams*; Levy, "Incomes and Income Inequality"; Newman, *Declining Fortunes*; and Schor, *The Overworked American*.

5. For discussion of the point that economic growth does not require inequality—and may perhaps call for more equality—see chapter 5.

6. We know that in statistical models of individual status attainment much, if not most, of the variance is unaccounted for. Of the explained variance, however, the bulk is due to social environment broadly understood. Also, we believe that much of the residual, unexplained variance is attributable to unmeasured social rather than personal factors.

7. Bock, *Human Nature Mythology*, p. 9.

8. Duster, review of *The Bell Curve*, 1995.

9. We are aware of three books on *The Bell Curve* that appeared since we began our project. Fraser's *The Bell Curve Wars* and Jacoby and Glauberman's *The Bell Curve Debate* are both largely collections of short reviews, some more substantive than others. While valuable, neither collection presents a sustained or systematic analysis. Dickens et al., *Does The Bell Curve Ring True?*, renalyzes the NLSY data as we do.

10. For example, at one point Herrnstein and Murray argue that teaching cannot raise IQ scores and at another they argue that scores can be elevated by "teaching to the test"; they claim that IQ is fixed early in life, but later they say that preschool programs can increase IQ scores, if only for a few years; they argue that "dysgenesis" (a decline in genetic "fitness" due to higher birthrates among the less fit) has been lowering American IQ, but later on that average academic skills have improved in recent decades; they contend at one point that equalizing intellectual abilities is nearly impossible to accomplish, but elsewhere that "leveling" has occurred in school skills (which they often use as a proxy for intelligence); and at many places they emphasize that small differences are important—arguing, for instance,

241

that a three-point drop in the average American IQ might create a nearly 10 percent rise in illegitimate births—whereas at others they dismiss small differences, the difference of a few points that, say, school enrichment programs make in IQ.

11. To determine how unequal the incomes of Americans would be if all adults had the same IQ, we took a high estimate of how much variation in income was attributable to variation in IQ and applied it to the distribution of household incomes in the United States in 1993. The largest correlation between the AFQT and an economic indicator is its correlation with the logarithm of hourly wages ($r = .31$). One of the fundamental theorems of statistics states that the proportion of variation in a dependent variable (wages in this case) left after equalizing everyone on an independent variable (the AFQT) is equal to $1 - r^2$. So equalizing everyone in the NLSY on the AFQT would reduce variation in hourly wages to ($1 - [.31]^2 =$) .90 of its original amount. We care less about ineqality in hourly wages in the NLSY than inequality in household incomes in the country as a whole. But we do not know the correlation between IQ and household income in the country as a whole, so we "borrow" the correlation between the AFQT and hourly wages as our high estimate of the correlation we seek. The solid line in figure 1.2 shows a log-normal distribution with the same mean, median, and variance as the actual distribution of household incomes in the United States for 1993. The dashed line shows a log-normal distribution with the same mean and median as the actual distribution of household incomes in the United States in 1993, but with a standard deviation only .9 as large as the actual one. This approach is generous to Herrnstein and Murray in the sense that we took the top estimate of r, .31. If we had taken the average of several appropriate r's, the results would show an even smaller consequence of equalizing IQs. Dickens et al., "Does *The Bell Curve* Ring True?" came up with a similar display and conclusion.

12. In a rejoinder to critics, Murray defended the low explained variance of *The Bell Curve* models by noting that few serious social scientists treat that statistic as the critical issue in their analyses; they focus instead on whether particular causes are statistically and substantively significant or not (Murray, "*The Bell Curve* and Its Critics"). This is true. But the reason social scientists usually downplay explained variance is that they recognize multiple causality—social effects are the result of many social causes. Therefore, finding a variable that adds significantly to a complex explanation, even if its explained variance is modest, is a success. Herrnstein and Murray, however, are engaged in a critically different enterprise. They argue that intelligence is *the* predominant cause of individual outcomes, virtually the only one with any force. Long before *The Bell Curve*, social scientists had listed IQ scores as *one* predictor among many predictors of individual success, but Herrnstein and Murray upped the ante and thus have to answer to a higher standard, their own claim that intelligence is *the* explanation. Their focus calls for examination of explained variance.

13. E.g., Goldberger and Manski, "Review Article: *The Bell Curve*"; Heckman, "Lessons from *The Bell Curve*"; David Levine, personal communication.

14. Murray and Herrnstein, "Race, Genes, and I.Q.—An Apologia," p. 34.

15. See *The Bell Curve*, p. 548 and p. 767, note 26, for Herrnstein and Murray's guaranteed annual income proposal.

16. Herrnstein and Murray argue that the Founding Fathers believed that natural differences in ability would lead to inequalities. But economic inequality was much less in their time than it is in ours (see chapter 5), and it is historically naive to use their acceptance of small inequalities to justify the large differences in life chances we see in America today.

CHAPTER 2

1. See, for example, Ceci, *On Intelligence*; Gardner, *Multiple Intelligences*; Sternberg, *The Triarchic Mind*; and Restak, *The Modular Brain*.

2. Kuhn, *The Structure of Scientific Revolutions*; Lakatos, *The Methodology of Scientific Research Programmes*.

3. Quoted by Tucker, *The Science and Politics of Racial Research*, p. 72.

4. Kranzler and Jensen, "The Nature of Psychometric *g*."

5. Jensen, "How Much Can We Boost IQ and Scholastic Achievement?"

6. So, for example, the psychometric defenders of *The Bell Curve* propose a vague list of attributes for what intelligence "involves," from the ability to think abstractly to learning from experience to "catching on" (Arvey et al., "Mainstream Science on Intelligence").

7. By combining two or more tests, each of which measures only one skill, we can build up to larger domains. For example, if we believe that a good football player will have size, speed, strength, and agility, we can use four tests, one for each domain, and combine the person's score on these subtests to obtain an overall score of "football ability."

8. The only way totally to rule out chance fluctuations in test scores would be to give someone an infinite number of tests. In most circumstances we have only a few test scores or only one. Therefore, we should not give much weight to fine distinctions between scorers. For example, in the verbal SAT a swing of thirty points up or down can be explained as random fluctuation. This implies that two students whose scores are as much as sixty points apart on the SAT should be treated as if they are equal, because the difference in scores could very well be the result of chance. Practically all test makers admit this error in their tests; it is not known, however, how many test users (e.g., admissions committees, employers) take this seriously when they evaluate candidates for positions. (The statistical information is courtesy of Nancy Wright, Educational Testing Service, November 1995; see also Educational Testing Service, *Handbook for the SAT Program*.)

9. Psychologist William Tucker (*The Science and Politics of Racial Research*, chap. 2) has described how Sir Francis Galton, one of the early statisticians and authors on intelligence, insisted that virtually all properties in the world *must* be distributed as a bell curve.

10. In particular, all scores on a normal distribution are completely determined by two quantities—its mean and its standard deviation. Other distributions require more information about the distribution.

11. The AFQT was designed to help the armed forces ascertain the readiness of high school juniors and seniors to join the military. The NLSY gave the test to high school graduates, college students, and college graduates as well. And because the test is really a test of school learning—as we discuss later—it is not surprising that so many test takers "hit the ceiling" of the test.

12. Herrnstein and Murray first converted each test taker's AFQT score to a centile (e.g., 99th, 98th, 97th). Then they used a standard table for the normal distribution to assign each test taker a Z-score equivalent to their percentile rank (see *The Bell Curve*, pp. 571–73).

13. Herrnstein and Murray partially address potential criticism of their normalizing procedure in an appendix section entitled "How Sensitive Are the Results to the Assumption that IQ is Normally Distributed?" (pp. 585–88). There, they try to show that, were the AFQT scores represented not as the normalized z-scores but as a series of dummy variables tied to centiles—in top 5 percent, next 20 percent, next 50 percent, next 20 percent, or bottom 5 percent—their statistical results would look pretty much the same. They present the regression of being in poverty on zAFQT, zSES, and zAge as an example, contrasting it to the same equation with four dummy variables in place of zAFQT. They claim it makes little difference. But, first, it *does* make a difference. The dummy-variable equation explains 20 percent less of the variance in the dependent variable than does the zAFQT equation (.0757 vs. .0942) despite the addition of three more variables; and the importance of zSES is *42 percent greater* in the dummy-variable version ($b = -.39$ vs. $-.27$.) Those changes alter the relative importance of AFQT and parental status in explaining poverty. Second and more important, the dummy variable construction again highlights the tails. Test takers at the 25th percentile and at the 75th percentile (roughly, IQs of 90 and 110) are treated as if they scored the same, but test takers at the 94th and 95th percentiles (roughly, IQs of 123 and 125) are treated as qualitatively different.

On the question of whether it makes a difference whether one uses the normalized scores or the raw scores, the answer is complex. The transformations increase the regression coefficients by only a small fraction. But the untransformed z-scores would run from -3 to only 1.6. Nobody would appear in the $+2$ standard deviation's "cognitive elite." Also, in the graphs Herrnstein and Murray use, the line for the AFQT would stop short of the right-hand, minimizing the visual impact and the implicit claim.

14. See, for example, Gould, *The Mismeasure of Man*; Tucker, *The Science and Politics of Racial Research*; various essays in Fraser (ed.), *The Bell Curve Wars*; and Jacoby and Glauberman (eds.), *The Bell Curve Debate*.

15. When people take a series of what psychometricians consider to be well-

constructed intelligence tests, those who score high on one test tend to score high on the others, and the people who score low on one tend to score low on the others. Thus, the tests are somewhat (but not perfectly) intercorrelated. Some tests are more correlated with one another than are other tests. These findings led English psychometrician and statistician Charles Spearman to hypothesize that an underlying factor common to the tests accounts for their intercorrelation. This underlying factor is more evident in some tests than in others, but it is generally present in all the tests. It is thus the *general* factor within the tests, the general factor of intelligence, called g for short. Given that "intelligence is what intelligence tests measure," this g is intelligence.

16. E.g., Jensen, *Straight Talk about Mental Tests*.

17. See, for example, Wagner, "Context Counts." Other studies have found that people who score well on "IQ tests" do not do especially well as leaders in stressful situations; they even tend to "babble" on. See Fiedler and Link, "Leader Intelligence."

18. Only one set of conditions can break the logical equivalence of predictive and criterion validity. If the expensive indicator has already been established as valid by some means other than predictive validity, then criterion validity and predictive validity are different. If so, criterion validity is a very useful tool, but if not, both predictive and criterion validity require two assumptions.

19. Jensen, *Straight Talk about Mental Tests*.

20. Thanks to meteorologist Tim Somers for clarifying this process.

21. One of a string of definitions proposed by a group of psychometricians writing in defense of *The Bell Curve* (Arvey et al., "Mainstream Science on Intelligence").

22. Kranzler and Jensen ("The Nature of Psychometric g"), for example, have published evidence suggesting the existence of four distinct and independent processes behind g. And they are unable to label or describe what those four processes are, so even this evidence does not move us much closer to designing successful public policy.

23. *The Bell Curve*, p. 580.

24. Arvey et al., "Mainstream Science on Intelligence."

25. The source for the questions is Bock and Moore, *Advantage and Disadvantage*, pp. 28–30. Because of confidentiality, the authors could not print the exact questions, but they assure us that the examples are quite close approximations.

26. One of the definitions of intelligence provided by the psychometricians defending *The Bell Curve*. See Arvey et al., "Mainstream Science on Intelligence."

27. Bock and Moore, *Advantage and Disadvantage*, p. 34; the alpha test is quoted in Lear, *Fables of Abundance*, p. 220.

28. Bock and Moore, *Advantage and Disadvantage*, p. 35.

29. Again, a definition from Arvey et al., "Mainstream Science on Intelligence."

30. Hacker, "Cast, Crime, and Precocity," p. 100.

31. Kohn and Schooler, "Occupational Experience," p. 101.

32. See, for example, Cornelius and Caspi, "Everyday Problem Solving," and discussions in chapters 3 and 7 of intelligence and aging.

33. Ceci and Roazzi, "The Effects of Context on Cognition."

34. The correlation of total SAT scores with college grades is .50; the correlation of high school grade-point average (GPA) is .48. Adding SAT scores to GPA raises the correlation to .59, an extra 1 percent of the variance (Educational Testing Service, *Handbook for the SAT Program*, p. 32).

35. $r = .50$ (ibid.).

36. For a group of selective universities, Herrnstein and Murray report that the gap between the average SAT scores of black freshmen and white freshmen ranges from 288 at Berkeley to 95 at Harvard.

37. Its standard deviation is 151 (26 above the national standard deviation).

38. Our twenty-student exercise is realistic in several important ways—the average of the scores, the difference between the averages of whites and blacks, and the overall dispersion of scores are close to national norms. It is unrealistic in one important way—it is very sensitive to the exact placement of the "black" cases. So we repeated the exercise starting with 20,000 applicants and admitting 10,000. As before, we assumed that 10 percent of the applicant pool is black. We added an additional element of realism by having 2,000 of the admittees choose to "decline admission," that is, we assumed that they did not enroll as freshmen even though they were admitted. We began by drawing a random sample of 18,000 white applicants from a normal distribution with a mean of 1000 and a standard deviation of 125 and 2,000 black applicants from a normal distribution with a mean of 875 and a standard deviation of 125. We ranked the students in this simulated applicant pool from high to low on SAT and "admitted" the first 10,000. The admitted applicants then were allowed to "enroll" or "decline admission" according to a probability determined by the formula

$$\text{prob(enroll)} = 1 - .0005 \times \text{sqrt(SAT)}.$$

Students with higher SAT scores were assigned a higher probability of declining admission because we assumed that they were more likely than students with lower scores to be admitted to more than one university. This formula resulted in 2,004 admittees declining admission. Because this system requires sampling, we repeated it ten times.

This more realistic system results in an average gap between white and African American freshmen of 85 points (averaged over the ten trials). It results in a freshman class that is only 7.2 percent black, however. As in the small example, the selection process is race-neutral, yet it results in a large gap between the freshmen of different groups.

39. To answer this question we reran our simulations—this time assigning a 50-point edge to the black students in the applicant pool. That is, we assigned each black student a place in the queue that she or he would get if her or his score were

50 points higher. This procedure increased the gap between the average test scores of white and black freshmen to 103 points (mainly by lowering the blacks' average). It is larger, but it is not 50 points larger than the 85-point gap we obtained using race-neutral selection. As an aside, we note that the "50-point edge" succeeds in raising blacks' share of the freshman class from 7 to 9 percent. In making more blacks eligible for admission, this affirmative action displaces 160 white applicants, which is less than 1 percent of the 18,000 white applicants.

40. *The Triarchic Mind*, p. 72.

41. Schraagen, "How Experts Solve a Novel Problem."

42. A. D. de Groot, *Thought and Choice in Chess*, as cited by Perkins, *Outsmarting IQ*.

43. Schoenfeld, *Mathematical Problem-Solving*.

44. One work that details the current state of knowledge within the information-processing paradigm is Sternberg's *The Triarchic Mind*.

45. Nisbett, "Race, IQ, and Scientism," p. 54.

CHAPTER 3

1. The basic source on the test is Bock and Moore, *Advantage and Disadvantage*.

2. Ibid., pp. 28–30.

3. The correlation of number right in the subtest with the final bell-curved score that Herrnstein and Murray used, zAFQT, was (whites only): arithmetic reasoning, $r = .90$; math knowledge, $r = .89$; words, $r = .85$; and paragraph comprehension, $r = .79$.

4. Recall the quotation from Bock and Moore, *Advantage and Disadvantage*, p. 34, cited in chapter 2, that the paragraph comprehension test leans heavily on general knowledge. Such general knowledge is conveyed in well-off homes as well as in schools.

5. Perhaps people with the same genetic endowments would score similarly on both kinds of tests. That is what the psychometricians, in effect, claim; but it is a claim based solely on assertion.

6. The "z" refers to "z-scored," meaning that all scores were readjusted so that the average would be zero and the standard deviation of the scores would be 1. One standard deviation is roughly the difference between someone who is exactly average, at the 50th percentile, with a z-score of 0, and someone who is at the 84th percentile, with a z-score of 1. A person two standard deviations up, z-score = 2, is at the 98th percentile. Someone at z-score = −1 is at the 16th percentile, and someone at z-score = −2 is at the 2nd percentile.

7. To be more complete: AFQT score increases modestly with age, and so does the probability of certain outcomes. For example, older respondents were likelier to have ever had a child out of wedlock or to have ever been jailed. Without controlling for age, it would appear that higher-scoring youths were *likelier* to have engaged in those negative behaviors!

8. Regressing the AFQT on the three predictors yields the following equation:

AFQT = −3.35 + .28 × (Years of Education at Test) + .53 × (Test Taker had Been in an Academic Track; yes = 1) − .17 × (Age, z-scored); R^2 = .37; n = 4,233.

See also Bock and Moore, *Advantage and Disadvantage*, on the issue of age.

9. Bock and Moore, *Advantage and Disadvantage*, systematically show in these data that age reduces scores within same-education groups, and they explicitly tie the finding to forgetting school lessons.

10. See, for example, Woodruff-Pak, "Aging and Intelligence," and Cornelius and Caspi, "Everyday Problem Solving in Adulthood and Old Age." A recent study suggests that lower test performance by people over seventy is largely explainable by their greater difficulties in seeing and hearing. See Lindenberger and Baltes, "Sensory Functioning and Intelligence in Old Age."

11. If exposure to education is poorly measured—and we have just argued that simply counting the number of years completed is substantively and statistically a poor measure—one will underestimate its effects. Also, the NLSY sample mixes together people who were still in school at the time of the test with people who had been out of school for years, further muddying the education effect.

12. Korenman and Winship, "A Reanalysis of *The Bell Curve*," p. 21. The result of the corrections yields an estimate that one year of school is worth 2.5 IQ points.

13. Using an instrument-variables approach that employed quarter of birth as an "instrument" for grades completed, Neal and Johnson found that each year of schooling in the NLSY sample increased the AFQT score about .23 standard deviations (versus .07 in Herrnstein and Murray's estimate), roughly equal to 3.5 IQ points. See Neal and Johnson, "The Role of Pre-Market Factors in Black-White Wage Differences," pp. 28–30.

14. Parcel and Menaghan, *Parents' Jobs and Children's Lives*, pp. 33–34.

15. Charles Murray is sensitive about the problematic correlation of years of education with AFQT scores. In response to attacks ("*The Bell Curve* and Its Critics"), Murray acknowledges that methodologists had criticized his approach to the issue and explains that he and Herrnstein wanted to avoid technicalities, and that he "look[s] forward to watching [his] colleagues apply those more sophisticated techniques" (p. 28). Still, he rests his claim that years of schooling do not account for life outcomes on *The Bell Curve*'s more primitive method of testing the AFQT-outcome effects. One of Herrnstein and Murray's methods was to replicate many of their findings looking only at respondents with a high school diploma. Among the difficulties with this procedure is the fact that, as we noted, one high school diploma does not equal another in exposure to quality academic instruction.

16. Eighteen percent of whites with four or more years of college were in the top 5 percent, compared with only 1.5 percent of those with no postsecondary education.

17. Bock and Moore, *Advantage and Disadvantage*, pp. 109–10, point out that

the math subtests were especially vulnerable to forgetting by test takers long out of school.

18. Psychometricians (Arvey et al., "Mainstream Science on Intelligence") state that an IQ of 70 to 75 is the border for mental retardation.

19. Here is the cross-classification of white respondents by (a) whether the interviewer thought they were "mentally handicapped" between 1989 and 1991, and (b) whether they scored below 70, "retarded," on the AFQT scores Herrnstein and Murray calculated:

		Handicapped?	
		No	*Yes*
Below 70?	No	4,275	32
	Yes	29	10

Of the thirty-nine scoring below 70, clearly retarded by psychometric standards, only 26 percent apparently remained handicapped when assessed several years later by interviewers. Of the forty-two assessed as handicapped by interviewers, only 24 percent had scored so low on the AFQT earlier. (It is possible that some brain damage occurred in the interim to the thirty-two cases above, but not likely. And it is not possible, if the theories of intellectual fixity are right, for the twenty-nine cases of retardation in 1980 to have been cured later.)

20. With four answer options per question, random responses would *average* twenty-six correct. Technically, random responding would yield a distribution of scores around twenty-six. About half the guessers would have gotten more than twenty-six right just by chance; so our estimate of their numbers is conservative. The point, however, is not to identify guessers—most of these exceptionally low scorers probably just stopped answering—but to underline how exceptional it must be to score so low.

21. Nine white test takers who scored 26 or fewer correct had IQ scores from their schools available. Three of them had scored above the 30th percentile in the earlier tests. Sixteen white test takers scored 27 or fewer correct and had earlier IQ scores available. Of those, three (almost one-fifth) had earlier IQ scores of 100 or higher. It is possible, of course, that between their earlier IQ tests and the AFQT some test takers had suffered brain damage. The same might be suggested for the test takers who were mentally disabled. But, the chances are small that many fell into this category or that they would have been able to take the AFQT in 1980.

22. Bock and Moore, *Advantaged and Disadvantaged*, appendix A; and Steve McClaski, Center for Human Resource Research, Ohio State, personal communication, February 8, 1995.

23. Milofsky, *Testers and Testing*.

24. See, e.g., Chapin, "The Relationship of Trait Anxiety"; and discussion of Claude Steele's work in chapter 8.

25. Technically, we are saying (a) that the problems we have noted with the AFQT are not random measurement errors, but systematically biased measurement

errors; and (b) that in regression equations, AFQT scores will remain robust even when controls for some of these other factors are introduced because the AFQT is "downstream" in the causal path. It summarizes a variety of influences.

CHAPTER 4

1. Murray, "*The Bell Curve* and Its Critics," p. 28.

2. Korenman and Winship, "A Reanalysis of *The Bell Curve*"; Dickens et al., "Does *The Bell Curve* Ring True?"; Heckman, "Lessons from *The Bell Curve*."

3. See, for instance, Jencks et al., *Inequality*.

4. Murray, "*The Bell Curve* and Its Critics."

5. For example, Duncan, "Inheritance of Poverty or Inheritance of Race?"; Duncan, Featherman, and Duncan, *Socioeconomic Background and Achievement*; Griliches and Mason, "Education, Earnings, and Ability"; and Jencks et al., *Inequality*, all make use of the AFQT itself.

6. The research community has been convinced enough to continually include ability measures on large peer-reviewed surveys such as the NLSY. The High School and Beyond study (Coleman and Hoffer, *Public and Private High Schools*) included extensive testing. The General Social Survey (Davis and Smith, *General Social Survey*) has routinely included a ten-item vocabulary test (Alwin, "Family of Origin"; Hauser and Carter, "*The Bell Curve* as a Study of Social Stratification") and in 1994 included a second measure, a short battery from the WAIS (see Davis and Smith, *General Social Survey*, appendix D).

7. See, e.g., Wegener, "Job Mobility and Social Ties"; Lin and Dumin, "Access to Occupations Through Social Ties."

8. In throwing down a challenge to his critics to come up with better measures of environmental influences, Murray suggests perhaps measuring the competency of a father or the moral standards of a family (Murray, "*The Bell Curve* and Its Critics," p. 29). This minimal interpretation of the environment pervades both *The Bell Curve* and much of the commentary about it. To the sociologist, the environment includes family dynamics but goes far beyond the family to the community and economy.

9. E.g., Massey and Denton, *American Apartheid*.

10. These are standard findings in sociological and economic research, e.g., Hogan and Kitagawa, "The Impact of Social Status"; Massey and Eggers, "The Ecology of Inequality"; Massey and Denton, *American Apartheid*; Crane, "The Epidemic Theory of Ghettos"; Rosenbaum and Popkin, "Why Don't Welfare Mothers Get Jobs?"; Wilson, "Social Theory and the Public Agenda"; Brooks-Gunn et al., "Do Neighborhoods Influence Child and Adolescent Development?"; and Brewster, "Race Differences in Sexual Activity among Adolescent Women."

11. When Herrnstein and Murray transformed the number of correct AFQT answers into a "normalized" AFQT index (see chapter 2), they used a transformation that produces a new variable, zAFQT, with a mean of zero and a standard deviation

of 1. But, as used in the multiple regression analyses, zAFQT has a standard deviation of .897 instead of 1, because (a) the NLSY tended to lose contact with the lowest-scoring people, so the variation in AFQT among those followed to 1990 was smaller than 1; and (b) the variation among whites in zAFQT score is less than among the sample as a whole. To correct for attenuated variation, we divided the logistic regression coefficient by .897, thus increasing it by 11.5 percent and giving Herrnstein and Murray the benefit of the doubt.

12. Among the white respondents, 14 percent were asked about neither their father's (or head of household's) occupation nor their parents' income because the respondents were living on their own. (Instead, these respondents were asked their own incomes and occupations.) An additional 3 percent failed to answer the occupation question and 7 percent failed to answer the income question although asked. Four and 6 percent, respectively, were missing information on mother's and father's education.

13. Arthur Goldberger reminded us that, in describing the construction of their index, Herrnstein and Murray said that for respondents who were missing components of the SES index, they assigned a score based on averaging the remaining components. Had they done that, there would have been error, but not such great error if the components were highly intercorrelated. However, we were unable to replicate their results with that procedure. We were able to replicate their results by assigning mean scores (0, since these are standardized measures) for missing data. Simply excluding a variable from an index is equivalent to assigning the mean since the mean of a normalized variable is 0 and not adding it to the index is the same thing as "adding 0."

14. Most notably, respondents without parental income information were older than the others ($r = .36$ between age and a dummy variable for missing information on parental income).

15. The procedure of assigning average income to respondents with missing data reduces its apparent influence, first, by introducing error—youth from poor and wealthy families are coded as coming from average families—and, second, by reducing the variance of the income measure, thereby reducing its statistical association with the effect variables.

16. We added a dummy variable (coded 1 if parental income was missing in either 1978 or 1979 and 0 if it was present in both). The coefficient for this dummy variable measures how far off the attribution of "average income" to missing cases is on average. Korenman and Winship ("A Reanalysis of *The Bell Curve*") use a much more sweeping adjustment. They take advantage of the total household design of the NLSY (all persons fourteen to twenty-two years old in a sampled household were included in the study) to match siblings. Through a "fixed effects" approach, they then capture all of the effects of origins, including neighborhood effects, mobility effects, and any other common links among siblings that might turn out to be important for poverty (or any of the other outcomes they explore). We rejected that approach because it is vulnerable to the claim that the main "shared variance" being

picked up by the fixed effects model is the shared genetic background of the siblings. Even though we doubt that such a thing is important for complex outcomes like poverty, fertility, and incarceration, it is all too clear that Herrnstein and Murray do not share our skepticism. Indeed they clearly favor the genetic interpretation, preferring it even for explaining the correlations between parental education and most of the outcomes they investigate. Our approach does not run the risk of picking up these kinds of purported genetic effects.

17. In one study, the correlation between what youths report their parents' income to be and what their parents report (which itself may be less than perfectly accurate, of course) was .82. This implies that correlations between reports of parental income and other variables will be about 18 percent too low. See Hauser et al., "The Intergenerational Transmission of Income Status."

18. Korenman and Winship, "A Reanalysis of *The Bell Curve*."

19. Parents' education is overweighted in the index because the two components, mother's and father's years of schooling, have much less missing data than do the other two components, parental income and occupation. Herrnstein and Murray assigned those with missing data the average score. This move compresses variation. So the education measures were compressed less than the occupation or income. The result is that variation in parental education contributes more to the variation in the total index than the putative 50 percent the authors imply. This technical error compounds a substantive one: Parents' education has no effect on the probability of poverty, while parental income does. So the index's construction—putting these measures together—deflates the apparent effect of parental SES.

20. Downey, "When Bigger Is Not Better"; Duncan, Featherman, and Duncan, *Socioeconomic Background and Achievement*; Featherman and Hauser, *Opportunity and Change*; DiPrete and Grusky, "Structure and Trend in the Process of Social Stratification."

21. Korenman and Winship, "Reanalysis of *The Bell Curve*."

22. It is not a perfect copy because *The Bell Curve* left out a few pieces of technical information we needed in order to follow its procedures exactly, but it is effectively the same result.

23. We also found that logging income reduced its effectiveness as a predictor, so we used income unlogged.

24. The technical reason we refer to here is the differential reliability of the AFQT and the parental home environment measures, as discussed in the earlier section on errors.

25. Murray, "*The Bell Curve* and Its Critics," pp. 28–29.

26. "Reanalysis of *The Bell Curve*," p. 22; Dickens et al., *Does the Bell Curve Ring True?*

27. In this section we build up from Herrnstein and Murray's statistical analysis to a more complete one. We start by completing our picture of respondents' backgrounds by adding aspects of the places they grew up in to the model. We follow this by adding education, then gender (two of the variables most analysts would start

with). We then consider people's contemporary environments. Finally, we add the current family circumstances.

28. See, e.g., Lichter et al., "Local Marriage Markets"; and Lichter et al., "Race and the Retreat from Marriage."

29. See, e.g., Myers and Wolch, "The Polarization of Housing Status."

30. Wilson, *The Truly Disadvantaged*; Massey and Denton, *American Apartheid*.

31. On concentration effects, see, for example, Brewster, "Race Differences in Sexual Activity"; Crane, "The Epidemic Theory of Ghettos"; and Hirschl and Rank, "The Effect of Population Density on Welfare Participation." See also Wacquant and Wilson, "Poverty, Joblessness, and the Social Transformation of the Inner City"; Wilson, "Social Theory and the Public Agenda"; Massey and Denton, *American Apartheid*; and Anderson, *Streetwise*. For reservations on concentration effects, see, e.g., Waters and Eschenbach, "Immigration and Ethnic and Racial Inequality," p. 428.

32. Chapter 7 discusses how schools make a difference.

33. School administrators, not respondents, provided these data, so they are of good quality. Each measure was recoded to have a mean of 0 and a standard deviation of 1 (the shape of the distribution was not "normalized"). Unfortunately, data are missing for 22 percent of the sample. The cases with missing school data are not a random sample of the total, so we gave the missing cases a score of 0 (average), and added a "dummy" variable coded 1 for missing cases and 0 otherwise. The value of the coefficient for the dummy variable indicates how much these cases differ from the average value we imputed to them.

34. See, e.g., Coleman and Hoffer, *Public and Private High Schools*.

35. We do not advocate "weighing" effects this way as a general practice. However, since Herrnstein and Murray have defined the terms of debate as being over the relative weight of test scores and environmental measures, we present our results in this way.

36. For this comparison we quantify "disadvantaged school" as 1.5 standard deviations below average and "advantaged school" as 1.5 standard deviations above average.

37. At one standard deviation below average on either AFQT scores or family origins, the poverty rate is about 27 percent for respondents from poorly-off schools and about 10 percent for those from well-off schools.

38. On social capital in schools, see Coleman and Hoffer, *Public and Private High Schools*; on part-time jobs and opportunities, see Hotz et al., "The Returns to Early Work Experience"; on marriage markets, see Wilson, *The Truly Disadvantaged*; Mare, "Five Decades of Educational Assortative Marriage"; Lichter et al., "Local Marriage Markets"; and Lichter et al., "Race and the Retreat from Marriage."

39. A classic reference on the subject is Jencks et al., *Inequality*. See DiPrete and Grusky, "Structure and Trend in the Process of Social Stratification."

40. The NLSY recorded 1990 unemployment rates for the local labor market in six general categories. We assigned mean values for those categories.

41. See appendix 2, table 2A, model A6.

42. In their own analysis (pp. 601–602), the AFQT correlated with being married only among those with a high school diploma, and at that very weakly.

43. Bianchi, "Changing Economic Roles of Men and Women."

44. The dependent variable in this analysis is whether the respondent was ever interviewed in jail during the 1980s. Aside from the other errors Herrnstein and Murray made, which we discussed earlier, they included men who were in jail *at the time* they took the AFQT, which raises the problem of whether their low AFQT scores caused jailing or vice-versa. Therefore, we restricted the dependent variable to incarceration *after* taking the test.

45. Korenman and Winship, "A Reanalysis of *The Bell Curve*," table 5b and appendix 2.

46. As in the case of jailed men, we restricted the analysis to whether or not women had a child out of wedlock *after* having taken the AFQT and after the other measures were taken as well.

47. The conclusion holds even if the education measures—years completed at time of AFQT, and academic track or not—are dropped from the equation.

48. Korenman and Winship, "A Reanalysis of *The Bell Curve*," appendix 2.

49. Calculations by Michael Hout from biographies of "Forbes 400" in *Forbes*, October 23, 1994.

50. Sullivan et al., *As We Forgive Our Debtors*.

51. Income statistics do not include "paper gains" like appreciation on the value of a house. And in surveys like the NLSY, income usually does not include the value of noncash benefits that people get from their employer, for example, the portion of health insurance premiums that the employer pays.

52. Bane and Ellwood, *Welfare Reform*.

53. Estimated from U.S. Bureau of the Census, *Statistical Abstract 1995*, table 740. This estimate includes men with no income at all.

54. The latest statistics indicate that in 1991 only 77 percent of divorced, custodial mothers received child support payments, and those payments amounted to only 16 percent of their income (ibid., table 617). In 1989, only 6 percent of divorced women were supposed to be receiving alimony payments (calculated from *Statistical Abstract 1994*, table 605).

55. Calculated from ibid., tables 677, 594, and 609, and U.S. Bureau of the Census, *Historical Statistics of the United States*, p. 356.

56. Both the U.S. Census Bureau and Statistics Canada agree that there are about twenty-nine million border crossings a year either way. The estimate from the PSID is that fifteen million Americans escape poverty every year. Since the number of poor is growing slightly, perhaps sixteen million enter poverty a year. The total is thirty-one million poverty-line crossings a year.

57. U.S. Bureau of the Census, "Poverty—Long and Short Term."

58. For the lower graph in this figure, we repeated the exercise we used to calculate figure 1.2 but examined how much inequality would remain if we equalized environmental factors. Using the NLSY, we took the multiple correlation between log wages and our environmental variables—parental home environment, adolescent community environment, and adult community environment—and gender, controlling for the AFQT, as our estimate of R. That multiple correlation, .61, is twice the correlation between the AFQT and log wages. Again, we "borrow" this correlation with wages in the NLSY as our estimate of the association between social environment and household income in the country as a whole. The calculation is $1-(.61)^2 = .63$, so 63 percent of the inequality in household income would remain if everyone had the same social environment and gender made no difference.

CHAPTER 5

1. Crystal, *In Search of Excess*, p. 27; *San Francisco Chronicle*, March 1, 1996, p. A12. Crystal carefully includes indirect compensation in these calculations.

2. *In Search*, chap. 13.

3. In 1989 the wealthiest 1 percent of American households owned 39 percent of all the household wealth in the nation. They owned almost half of all the financial wealth (largely liquid wealth, such as savings, stocks, and bonds, but not real estate). See Wolff, *Top Heavy*, p. 7.

4. The gini coefficient for major league salaries increased from roughly the .35 range to the .50 range from 1965–74 to 1986–90 (Fort, "Pay and Performance"). In 1976, before free agency, the highest-paid player, Johnny Bench, made about five times the median salary; by 1992 and 1993, the highest-paid player, Bobby Bonilla, made fourteen times the median salary (Joe Harder, personal communication, February 16, 1995).

5. More accurately, Charles Murray and others have argued that government does affect these processes, but only in ways that interfere with rewarding talent. With less government, they suggest, we should see even more inequality.

6. The best sources are Williamson and Lindert, *American Inequality*, and "Long-Term Trends in American Wealth Inequality." See also Soltow, "Wealth and Income Distribution"; Story, "The Aristocracy of Inherited Wealth"; Lebergott, *The American Economy*; Bruchey, *Enterprise*; Kulikoff, "The Transition to Capitalism in Rural America"; Fogel, "Nutrition and the Decline in Mortality"; and Wolff, *Top Heavy*. A recent reevaluation of the data is Shamas, "A New Look at Long-term Trends in Wealth Inequality."

7. An excellent survey of this process is Blumin's analysis of Philadelphia, *The Emergence of the Middle Class*. For other studies of increasing inequality in the colonial and early Republican eras, see, for example, Wolf, *Urban Village*; Greven, *Four Generations*; Johnson, *A Shopkeeper's Millennium*.

8. Shamas, "A New Look at Long-Term Trends," argues that this standard ac-

count underestimates inequality in the colonial era, because it ignores the wealth British residents held in the colonies and it treats households as units of analysis rather than individuals. Making her corrections, which are debatable, would dampen somewhat the apparent increase in inequality during the nineteenth century. It would not change our basic claim, that the extent of inequality fluctuated notably across time.

9. To take advantage of the offer, settlers had to have enough cash to travel, put down a filing fee, and set up farming. The most economically viable land was granted to the railroads who then resold it. Also, speculators quickly cornered much of the land. Ultimately, two-thirds of homesteaders failed.

10. Danziger and Gottschalk, "Introduction."

11. Lebergott, *The American Economy*, p. 508. The statistic refers to the proportion of husband-wife families with low incomes, not including aid-in-kind.

12. Soltow, "Wealth and Income Distribution," emphasizes some of these other dimensions of equality.

13. Initial social security legislation excluded agricultural and household service workers, who were heavily black. See Quadagno, *The Color of Welfare*.

14. U.S. Bureau of the Census, *Historical Statistics*, pp. 19–20.

15. Ford, *Work, Organization, and Power*, p. 4.

16. One early and notable exception to the conservative deniers was Kevin Phillips. By 1995 other conservative voices had conceded the point—e.g., George Will ("What's Behind Income Disparity").

17. These numbers refer to households, including single-person households, which increase the proportion of all households during the period. However, family-only data show the same basic pattern—a shrinking gap to about 1970 and a widening gap afterward. See U.S. Bureau of Census, *Historical Statistics*, p. 301, and Ryscavage, "A Surge in Growing Income Inequality?" p. 54. The 1994 figures are from the Census Bureau's World Wide Web page at www.census.gov/ftp/pub. See, on families, U.S. Bureau of Census, Current Population Reports P-60, no. 174 and no. 184. Another measure of inequality is the gini coefficient. By 1970 it had dropped to .39. By 1993 it had risen to .45 (Ryscavage, "A Surge in Growing Income Inequality?").

18. A short reading list includes Danziger and Gottschalk (eds.), *Uneven Tides*; U.S. Bureau of the Census, "Studies in the Distribution of Income"; Bluestone, "The Inequality Express"; Cutler and Katz, "Rising Inequality?"; Levy, *Dollars and Dreams*; Levy and Michel, *The Economic Future of American Families*; Levy and Murnane, "U.S. Earnings Levels and Earnings Inequality"; Mishel and Frankel, "Hard Times for Working America"; Duncan and Rodgers, "Has Children's Poverty Become More Persistent?"; Lichter and Eggebeen, "Rich Kids, Poor Kids"; U.S. Bureau of the Census, "Workers with Low Earnings."

19. Research of Greg Duncan and Timothy Smeeding quoted in Bradsher, "America's Opportunity Gap." See, also, Browne, "The Baby Boom and Trends in Poverty," who concludes that "every successive generation of family heads born

throughout the baby boom has faced an increasingly greater chance of being poor" (p. 1071).

20. Wolff, *Top Heavy*, pp. 7–8. Weicher, "Changes in the Distribution of Wealth," suggests that the evidence is murkier and the change less sharp from 1983 to 1989. Wolff, "How the Pie Is Sliced," however, shows that the trend toward greater inequality continued at least to 1992.

21. The data are from U.S. Bureau of the Census, "Workers with Low Earnings," and Jack McNeil, U.S. Bureau of the Census, personal communication, November 1995.

22. Between 1977 and 1988, 60 percent of men who turned 30 were already earning about $23,000 (in 1993 dollars), but between 1988 and 1992, that had dropped to 42 percent (research of Greg Duncan and Timothy Smeeding quoted in Bradsher, "America's Opportunity Gap").

23. In 1970, 95.9 percent of men aged 25–54 were in the civilian labor force; in 1994, 91.8 percent were (calculated from U.S. Bureau of the Census, *Statistical Abstract 1995*, table 627).

24. Levy, *Dollars and Dreams*, p. 78.

25. U.S. Bureau of the Census, "Income and Job Mobility in the Early 1990's."

26. Marshall, "Wage Picture," reports on the work of a commission headed by Stanford economist Michael Boskin. See also Samuelson, "Wages, Prices, and Profits." For counter-arguments suggesting that poverty has been underestimated, see, for example, Cutler and Katz, "Rising Inequality?"

27. See note 23.

28. For instance, fewer Americans can afford to buy homes. According to census calculations, the proportion of married couples who could afford a "modestly price" home in their area—including couples who currently did own a home—dropped from 60.4 percent to 57.6 percent from 1984 to 1991 (U.S. Bureau of the Census, "Housing Affordability"). In 1991 current homeowners carried more debt on their homes than a decade earlier (U.S. Bureau of the Census, "Debt"; see also Karoly, "The Trend in Inequality," p. 46; U.S. Bureau of the Census, "Workers with Low Earnings"; Levy, *Dollars and Dreams*; and Newman, *Declining Fortunes*.) In 1992 *Consumer Reports* ("Has Our Living Standard Stalled?") made the following calculations. They took the actual prices of specific goods in different years and computed how much work time, at the average hourly wage of the year, it would take to buy them. For the small items, the cost in work time had generally dropped, but for the big items, it had increased. Here are some examples:

	1972	*1992*	*Change*
First-class postage	1.6 min.	1.6 min.	0
Phone call, New York to Los Angeles	23.5	4.3	−82%
Gasoline, 1 gallon	5.2	6.4	+23
Chicken, 1 lb., whole, cut	8.7	7.4	−15

(*continued*)

	1972	1992	Change
Milk, half-gallon	9.6 min.	7.8 min.	−19%
Film, Kodak, 35mm, color	37.9	27.2	−28
Record/CD album	2.2 hours	1.6 hours	−27
Television, RCA, 19 inches	121.6	21.7	−82
Washing machine, Sears, midprice	52.7	37.8	−28
Mattress, Simmons, with box	59.5	60.7	+02
Income taxes, federal, including social security	48.3 days	49.0 days	+01
Child delivery	37.2	62.2	+67
College, public, 1 year	64.1	99.2	+94
Car, average, new	131.0	197.8	+51
House, 3 bedroom, Matawan, N.J.	1,330.7	1,777.3	+34

Items that take minutes or hours of work tended to cost less in labor in 1992 than in 1972. But items that cost days of work (except federal income taxes) cost much more in labor. (Official statistics are consistent. The dollar cost of food went up 2.9-fold between 1973 and 1993, but shelter went up 3.8-fold and medical costs 5.2-fold. See U.S. Bureau of the Census, *Statistical Abstract 1994*, table 747.) The big items do not make up a large part of the market basket because they are rarely purchased. But they do make up a large part of the middle-class way of life. Their increasing cost helps explain the middle-class angst. For more on evaluation of the CPI, see Baker, "The Inflated Case against the CPI."

29. In 1975, 45 percent of women *with husbands present* and children under 18 at home were in the labor force; in 1993, the percentage was 69 percent (U.S. Bureau of the Census, *Statistical Abstract 1995*, table 639).

30. See, for example, Lichter and Eggebeen, "Rich Kids, Poor Kids"; and Browne, "The Baby Boom and Trends in Poverty, 1967–1987"; Karoly and Burtless, "Demographic Change."

31. Wolff, *Top Heavy*; Wolff, "How the Pie Is Sliced." But see Weicher, "Changes in the Distribution of Wealth."

32. Bluestone, "The Inequality Express," p. 82.

33. Ibid. See also Gottschalk and Smeeding, "Cross-National Comparisons of Levels and Trends in Inequality," pp. 16–33, for a review of a variety of explanations.

34. Economists making the case for a technological explanation include Krugman, "Technology's Revenge"; Murnane et al., "The Growing Importance"; and Krueger, "How Computers Have Changed the Wage Structure." Dissenters to this view include Howell, "The Skills Myth."

35. See, e.g., Roos, "Hot Metal to Electronic Composition." Also, Bureau of Labor projections suggest that the jobs that are growing are the ones that involve less-technical skills, such as health aides and childcare workers (Krugman, "Technology's Revenge," p. 59).

36. See, for example, Gottschalk and Joyce, "The Impact of Technological Change."

37. *New York Times*, April 16, 1994: 25. A couple of readers have noted that Eisner may have cashed in on years of Disney growth by selling stock options. Whether he "deserved" his $203 million by American standards, the striking fact is that CEO remuneration has remained seemingly immune to the pressures that have hit the average working American—not to mention the scale of compensation.

38. Atkinson et al., "Income Distribution"; Gottschalk and Smeeding, "Cross-National Comparisons"; and Gottschalk and Joyce, "The Impact of Technological Change."

39. Human capital economists might argue that the increasing variation in income simply reflects increasing variation in human abilities. But that would ignore the obvious fact that wages are affected by many external considerations, such as competition, having little to do with individual abilities. Also, the key component to "human capital," education, has generally been getting more homogenous, as a higher proportion of Americans go to college. And increasing inequality has occurred within most sectors and occupations.

40. The recent statistics are from *Statistical Abstract 1995*, table 627. See U.S. Bureau of the Census, "Studies in the Distribution of Income"; Haber and Gratton, *Old Age and the Search for Security*; Duncan and Smith, "The Rising Affluence of the Elderly"; and Achenbaum, "Old Age."

41. In 1993 the median net worth of households headed by a person over 64 was $86,324; the national median was $37,587 (U.S. Bureau of the Census, "Asset Ownership of Households: 1993").

42. *Statistical Abstract, 1994*, table 1220, and *1995*, table 1228.

43. *Statistical Abstract 1995*, table 169. In 1993, 15.3 percent of all Americans were without health insurance for the *whole year*, up from 14.4 percent in 1991. While 26.8 percent of 18-to-24-year-olds lacked coverage, only 1.2 percent of older people did (U.S. Bureau of the Census, "Health Insurance Coverage—1993").

44. Of men 25–54 who changed from one full-time job to another in the 1991–1993 period, 49 percent had had health insurance in their old job, but only 32 percent had it in their new job (U.S. Bureau of the Census, "Dynamics of Economic Well-Being").

45. Williams and Collins, "US Socioeconomic and Racial Differences in Health," pp. 351ff.

46. More specifically, the data are for male workers "aged 25–54, living in households with no self-employment income. Wages are net of employer contributions to social insurance (payroll taxes), but gross of employee payroll taxes" (Gottschalk and Smeeding, "Cross-national Comparisons of Levels and Trends in Inequality," table 1). The graph for women only would show Canada with the widest inequality, followed closely by the United States and the other nations far behind. Note that the ratios for low earners are transformations of Gottschalk and

Smeeding's estimates "p10," the ratio of the 10th percentile to the median × 100. Our figure is 100/(p10).

47. Freeman, "How Labor Fares in Advanced Economies."

48. Freeman and Katz, "Rising Wage Inequality"; Card and Freeman, "Small Differences that Matter"; Atkinson et al., "Income Distribution in Advanced Economies"; Gottschalk and Smeeding, "Cross-National Comparisons." In some European countries, wage agreements are negotiated between unions and management. Higher education, traditionally restricted in Europe to the few, expanded in the 1980s, lowering the wage premium graduates could claim.

49. Gottschalk and Smeeding, "Cross-National Comparisons," table 3. See also O'Higgins et al., "Income Distribution and Redistribution"; and other contributions in Smeeding et al. (eds.), *Poverty, Inequality, and Income Distribution.*

50. The United States is more unequal in household income than countries such as Austria, Finland, and New Zealand, too. The one exception is that Ireland (pop. 3.4 million) had the highest gap between rich and middle, slightly above that of the United States. Ireland was less unequal below the median than the United States.

51. Gottschalk and Smeeding, "Cross-National Comparisons," use posttax and posttransfer "household income per equivalent adult, using an 'intermediate' equivalence scale with E equal to 0.5" (p. 34). Note that the ratios for low-income households are calculated from the Gottschalk and Smeeding estimates, "p10," the ratio of the 10th percentile to the median × 100. Our figure is 100/(p10).

52. Ireland, not shown, has a rich-to-poor ratio of 4.2.

53. Smeeding et al. (eds.), *Poverty, Inequality, and Income Distribution*; McFate et al., "Markets and States."

54. Wolff, "How the Pie Is Sliced," p. 60.

55. See various contributions to Smeeding et al (eds.), *Poverty, Inequality, and Income Distribution.*

56. Smeeding et al., "Patterns of Income and Poverty." Similarly, American infant mortality rates for whites only exceed those of many other Western nations.

57. In the early 1920s, the richest 1 percent of families owned 37 percent of the wealth in America, but 59 percent in Great Britain—Wolff, "How the Pie Is Sliced," p. 60.

58. Smith, "Social Inequality." See, also Kelley and Evans, "The Legitimation of Inequality"; Rainwater, *What Money Buys*; McCloskey and Zaller, *The American Ethos*; Verba and Oren, *Equality in America*; and Hochschild, *What's Fair?* Recent work by James Kluegel shows complex international differences, but Americans remain distinct in resisting either minimum guaranteed incomes or ceilings on incomes, and also in blaming the poor for their situation (see Kluegel and Miyano, "Justice Beliefs," and Kluegel et al., "Accounting for the Rich and the Poor").

59. Quoted by Bradsher, "America's Opportunity Gap."

60. See, for example, Thomas, "Rising Tide Lifts the Yachts," subtitled "The Gap between Rich and Poor Has Widened, but There Are Some Comforting

Twists," and Will, "What's Behind Income Disparity," who wrote: "the problem of increasing inequalities of wealth is not a problem we will pay just any price to remedy, and may not be a problem at all. . . . [I]ncreasingly unequal social rewards can conduce to a more truly egalitarian society."

61. Persson and Tabellini, "Is Inequality Harmful for Growth?"; Chang, "Income Inequality and Economic Growth"; and especially Clarke, "More Evidence on Income Distribution and Growth." See also, Lindert, "Social Spending."

62. See Osberg, *Economic Inequality in the United States.* Wolff ("How the Pie Is Sliced," p. 64) also notes: "Analyses of historical data on the U.S. as well as comparative international studies confirm a positive association between equality and growth."

63. Between 1970 and 1990, American per capita gross domestic product grew by 443 percent; that of the European Union grew 526 percent; and that of Japan, 682 percent. This comparison adjusts for the different costs of living in the countries so that it is a rough measure of changes in the standard of living. U.S. Bureau of Census, *Statistical Abstract 1994*, table 1370.

64. Blank, "Does a Larger Social Safety Net Mean Less Economic Flexibility?"

65. Cowherd and Levine, "Product Quality and Pay Equity."

66. Persson and Tabellini's explanation ("Is Inequality Harmful?") for their results is that in societies with greater earnings equality, there is less political pressure for government redistribution; such redistribution impairs growth. However, their evidence for the explanation is thin, and Clarke's results ("More Evidence") are inconsistent with that argument. Chang ("Income Inequality") suggests that with more equality, lower-income families could make longer-term investment decisions. In any event, the statistical results suggest that government intervention on behalf of equality in the market, rather than after the market, would be beneficial.

67. See, for example, Porter, *Capital Choices*, pp. 53ff.

68. Quoted by Trachtenberg, *The Incorporation of America*, p. 75.

69. See, for example, Rainwater, *What Money Buys*; Kluegel and Smith, "Beliefs about Stratification."

70. Ganzeboom et al., "Comparative Intergenerational Stratification Research."

CHAPTER 6

1. U.S. Bureau of the Census, "Measuring the Effect of Benefits and Taxes," table 2, definitions 2 and 13. See also Smeeding, "Why the U.S. Antipoverty System Doesn't Work Very Well"; Smolensky et al., "The Declining Significance of Age in the United States." Another view of these effects is to examine the "gini coefficient," a measure of inequality. The coefficient runs from 0 to 1. The higher the number, the greater the inequality. The effects of government on this measure in 1992 are displayed below: One can see the great difference government makes for the elderly—a drop of 37 percent in the coefficient of inequality, from .649 to

.409—compared with the minor effect it has on families with two parents and children—a drop of 15 percent, from .368 to .312.

GINI COEFFICIENTS

	All Households	Married Couples with Children	Households with Member Over 64
Market Income Inequality	.497	.368	.649
And the effect of taxes	.471	.340	.616
And non–means-tested transfers, plus Medicare	.404	.326	.418
And means-tested transfers	.385	.312	.409

Source: U.S. Bureau of the Census, "Measuring the Effects"; income definitions 4, 8, 10, and 14.

2. Smeeding, "Why the U.S. Antipoverty System Doesn't Work"; Gramlich, et al., "Growing Inequality in the 1980s." The latter estimate that 62 percent of the increase in inequality in the 1980s could be attributed to pretax, pretransfers economic changes, but 38 percent to the reduced effectiveness of taxes and transfers to correct for market inequality (p. 237).

3. The spending estimate is of "Cash and noncash benefits for persons of limited income," table 577 of U.S. Bureau of the Census, *Statistical Abstract 1994*. (The 1995 *Abstract* has the same data.) Other numbers come from tables 727, 464, 733, 703, and 510. The number of low-income Americans used for the per capita estimates is the number whose individual or family incomes were below 125 percent of the poverty line in 1992—49.2 million people.

4. Burtless, "Public Spending on the Poor."

5. See Garfinkel and McLanahan, *Single Mothers and Their Children*; McLanahan and Sandefur, *Growing up with a Single Parent*; Bumpass, "What Is Happening to the Family?"; Hoynes, "Does Welfare Play Any Role in Women's Family Decisions?"; Moffitt, "Incentive Effects of the U.S. Welfare System"; Schultz, "Marital Status and Fertility."

6. More specifically, Rainwater and Smeeding, "Doing Poorly," measured the "real" disposable incomes of households in which the children resided in the late 1980s and early 1990s. They took household incomes and adjusted them, first, for the size of the households in which they lived and, second, for the different purchasing power in each nation (i.e., controlling the different costs of basic commodities across nations). The second adjustment was to enable them to compare living standards. They calculated incomes for children *before* taking into account any government taxes or transfer payments of cash or near-cash equivalents, and then the incomes *after* such government actions. (Not included are difficult-to-price services such as health care and day care.)

7. The following lists the percentage of children in poverty by number of parents:

	All Children		With Two Parents		With 1 Parent	
	Before	After	Before	After	Before	After
United States	26	22	14	11	70	60
Australia	20	14	12	8	73	56
Canada	22	14	15	7	68	50
United Kingdom	30	10	22	8	76	19
Italy	12	10	11	10	32	14
W.Germany	9	7	5	2	44	4
France	25	6	23	5	56	23
Netherlands	14	6	8	3	80	40

The numbers are from Rainwater and Smeeding, "Doing Poorly," table A-2. See also Garfinkel and McLanahan, "Single-Mother Families and Government Policy."

8. Rainwater and Smeeding, "Doing Poorly," table 1. Susan Mayer ("A Comparison of Poverty and Living Conditions") reports, based on examining surveys of consumption, that the American poor are no worse off than the poor in three other Western nations, even though their incomes are lower. She suggests that the American poor find ways, licit and illicit, of making up the difference. One way is by depending more heavily on family, friends, and charity. Questions remain about the methods of the study; also, many comparisons and many items are not included, such as child care and physical security. A deeper point is that the concept of poverty in that study simply considers it to be the lack of particular goods and services; poverty is more profoundly a matter of participation in the wider society—an issue of relativity (see, e.g., Rainwater, *What Money Buys*). In any event, a higher proportion of Americans are, in the end, poor than elsewhere, both materially and in the sense of cultural isolation.

9. Americans believe that they spend huge sums to support single mothers and their children. In fact, the entire budget for AFDC, the major program that supports poor women and children, was 1.6 percent of the federal budget in 1993. (Abolishing the entire $21 billion AFDC budget for the 1994–95 fiscal year would not have paid one $31 billion *monthly* premium for the interest on the national debt.) AFDC families typically receive other benefits, such as food stamps and Medicare, but AFDC is the cash allowance program that middle Americans most resent.

10. U.S. Bureau of the Census, *Statistical Abstract 1995*, tables 609, 762.

11. See, for example, Brock, *Welfare, Democracy, and the New Deal*; Katz, *In the Shadow of the Poor House*; and Trattner, *From Poor Law to Welfare State*.

12. Sources on the EITC include Holtzblatt et al., "Promoting Work through the EITC"; Scholz, "The Earned Income Tax Credit"; and Lerner, "Making Work Pay."

13. Peterson, "GOP Seeking to Curb Tax Break for Poor"; Novack, "The Worm in the Apple"; "IRS Appears Successful in Efforts to Curb Fraud," *Wall Street Journal*, June 9, 1995, p. 16; "The War on Work," *Newsweek*, October 2, 1995, p. 66.

14. Mayer and Jencks "War on Poverty."

15. This calculation assumes the taxpayer is at a 34 percent marginal tax rate and is paying a mortgage interest rate of 9 percent.

16. Grigsby, "Housing Finance and Subsidies in the United States"; Coontz, *The Way We Never Were*, p. 87. Dreier and Atlas, "Housing Policy's Moment of Truth," p. 70, lists the following expenditures (in billions):

Mortgage interest and property tax deductions for homeowners—$64
HUD subsidies to public agencies, private developers, and landlords—$26
Tax breaks for investors in rental housing and mortgage bonds—$13
Military subsidies to house personnel—$10
Welfare payments—$7
Rural subsidies—$3

17. These mortgage guarantees are now administered primarily through Ginnie Mae and Freddie Mac rather than through the FHA and the VA. On the growth of homeownership in the twentieth century, see Chevan, "The Growth of Home Ownership: 1940–1980"; Tobey et al., "Moving Out and Settling In."

18. Jackson, *The Crabgrass Frontier*; Coontz, *The Way We Never Were*, p. 24. See also Schneiderman, "The Hidden Handout."

19. See Jackson, *The Crabgrass Frontier*; Coontz, *The Way We Never Were*, p. 77. On the GI Bill, see Chafe, *The Unfinished Journey*.

20. Jackson, *The Crabgrass Frontier*; Coontz, *The Way We Never Were*; Chafe, *The Unfinished Journey*.

21. *Business Week*, May 11, 1992, p. 20; Dreier and Atlas, "Housing Policy's Moment of Truth," p. 74.

22. For good overviews of housing policies in other countries, see van Valet (ed.), *International Handbook of Housing Policies*; Husttman and Fava (eds.), *Housing Needs and Policy Approaches*.

23. Some rent subsidies are available in the United States under section 8 of the 1974 Housing Act, but this program has been very ineffective. Eligibility requirements are much more stringent than in other countries, and because the program is extremely underfunded, only an exceedingly small proportion of those eligible actually receive subsidies.

24. Heisler, "Housing Policy and the Underclass"; and Headey, *Housing Policy in the Developed Economy*. Coontz (*The Way We Never Were*, p. 87) points out that in the United States publicly owned housing accounts for only 1 percent of the housing market, while 37 percent of housing is publicly owned in France and 46 percent in England.

25. Jackson, "Race, Ethnicity, and Real Estate Appraisal" and "The Spacial Di-

mensions of Social Control"; Sugrue, "The Structures of Urban Poverty"; Massey and Denton, *American Apartheid.*

26. Duster, "The Advantages of White Males." See also Abrams, "The Housing Problem."

27. Duster, "The Advantages of White Males"; Quadagno, *The Color of Welfare*, p. 91.

28. Oliver and Shapiro, *Black Wealth/White Wealth*, tables 5.1, 5.2.

29. Wolfe, "Reform of Health Care," pp. 253–54. Americans spent $631 per person on health care in 1960 and $2,566 in 1990 (both in 1990 dollars), while health expenditures as share of GNP increased from 5.3 percent of GNP in 1960 to 12.2 percent in 1990, with conservative estimates that we will be spending at least 15 percent of GNP on health care by the year 2000.

30. Ibid., p. 254.

31. U. S. Bureau of the Census, *Population Profile*, p. 37.

32. Wolfe, "Reform of Health Care," p. 255.

33. Ibid., p. 265.

34. Wolfe has broken down the percent uninsured all year as follows:

Age	
Under 6	9.7
6–18	11.6
19–24	20.3
25–54	10.4
55–64	8.6
All ages under 65	11.4
By Income Needs	
Below the poverty line (PL)	27.5
Poverty line to $1.25 \times$ PL	26.6
$1.25 \times$ PL to $2 \times$ PL	19.5
$2 \times$ PL to $4 \times$ PL	8.1
More than $4 \times$ PL	3.5
Ethnic/Racial Group	
White	8.8
Black	16.2
Hispanic	25.3

35. Wolfe's breakdown is as follows:

Income Needs by Decile	% Reporting Health Limitations	% Reporting Poor Health
1	21.4%	30.7%
2	16.2	19.0

(*continued*)

Income Needs by Decile	% Reporting Health Limitations	% Reporting Poor Health
3	12.5%	11.9%
4	10.3	10.4
5	8.6	9.3
6	6.8	4.9
7	6.4	4.4
8	5.3	2.3
9	5.7	4.3
10	6.6	2.9
Total	100.0%	100.0%

Calculations use data from the 1980 National Medical Care and Expenditure Survey.

36. Garfinkel and McLanahan, "Single-Mother Families, Economic Insecurity, and Government Policy," p. 210.

37. Under the EITC's complicated 1993 rules, a family with one child that earned about $9,000 a year would have gotten an additional refund worth about $1,500 if they had a second child. However, the 1993 EITC legislation provided no additional support for children beyond the first two. Also, the EITC pays *less* to families that have less (a second child would yield only $800 more if the family earned only $5,000). And the EITC was cut back in recent legislation.

38. U.S. Bureau of the Census, *Statistical Abstract 1995*, tables 523 and 524.

39. "Right, Left Call for Cuts to Corporate 'Aid,'" *San Francisco Chronicle*, March 7, 1995; "A Hard Look at Corporate 'Welfare,'" *New York Times*, March 7, 1995, p. C1. A longer list of what seem flamboyantly unnecessary expenditures can be found in one of the reports on which the news stories were based, Moore and Stansel, "Ending Corporate Welfare as We Know It." Highlights include 1994 expenditures of $140 million to build roads in national forests, thus subsidizing the removal of timber from federal lands by multimillion-dollar timber companies; low-interest Rural Electrification Administration loans that reduce the cost of running ski resorts in Colorado and gambling casinos in Las Vegas; and a U.S. Department of Agriculture Market Promotion Program that in 1991, at a cost of $110 million, advertised abroad such items as Sunkist oranges ($10 million), McDonald's Chicken McNuggets ($465,000), and American legend mink coats ($1.2 million), as well as the Pillsbury baked goods that made it into the newspapers (pp. 3–4).

40. Shapiro, *Cut-and-Invest*, pp. 17–18.

41. Ibid., pp. 19–20.

42. Moore and Stansel, "Ending Corporate Welfare as We Know It."

43. Shapiro, *Cut-and-Invest*, pp. 20–21.

44. Ibid., pp. 22–24.

45. Ibid., pp. 15–16. Shapiro assumes that any job protection for workers in

assisted companies is offset by job losses to other companies that must compete with the subsidized firm for workers and investment.

46. Phillips, *Boiling Point*, p. 28.

47. Ibid., pp. 43–44.

48. A more detailed picture can be found in the following data from Phillips, *Boiling Point*, p. 113, on the comparative benefits of the Tax Reform Act of 1986:

Income Bracket	Size of Tax Cut	Average 1989 Tax Savings per Return
Up to $10,000	11%	$ 37
$10,000–$20,000	6%	69
$20,000–$30,000	11%	300
$30,000–$40,000	11%	467
$40,000–$50,000	16%	1,000
$50,000–$75,000	16%	1,523
$75,000–$100,000	16%	3,034
$100,000–$200,000	22%	7,203
$200,000–$500,000	27%	24,603
$500,000–$1,000,000	34%	86,084
$1,000,000 or more	31%	281,033

49. Ibid., pp. 110, 113.

50. ibid., p. 117.

51. Ibid., p. 112.

52. Freeman and Katz, "Rising Wage Inequality," pp. 36–43.

53. We say "in large part" because some of the greater wage inequality in the United States is due to the relative decrease in the 1980s in the number of workers with college degrees—a point we return to later. Some economists would argue that all that government policy overseas has done is to advantage workers lucky enough to hold on to their jobs while increasing unemployment for others. To a certain extent, they are right: There does appear to be some correlation between rising wages and lowered inequality, on the one hand, and sluggish employment growth, on the other, implying a trade-off between equality and employment. Our response to this line of argument is twofold. First, the most careful study to date indicates that even after taking this trade-off into account, institutions play an important role in accounting for wage inequality (Freeman, *Working under Different Rules*). Second, even if this trade-off existed and was the total explanation for levels of wage inequality, it still would demonstrate the importance of social policy choices.

54. Blackburn et al., "Changes in Earnings Differentials in the 1980s"; Freeman, "How Much has De-unionization Contributed"; Card, "The Effect of Unions on the Distribution of Wages." We say "at least" because it does not take into account the ways that lower unionization rates decrease the pressure on nonunion employers to

pay higher wages and provide benefits (Freeman and Katz, "Rising Wage Inequality," p. 48).

55. Freeman and Medoff, *What Do Unions Do?*

56. Card and Freeman, "Small Differences that Matter," pp. 199, 210, 211.

57. Delays in holding elections are highly correlated with election losses by unions. On election delays, see Freeman and Medoff, *What Do Unions Do?*, pp. 233–39. On the tactics of management consultants, see Fantasia, *Cultures of Solidarity*.

58. Card and Freeman, "Small Differences that Matter," p. 199. Canadian figures are an average for the 1980s.

59. Lemieux, "Unions and Wage Inequality"; Card and Freeman, "Small Differences that Matter."

60. Kasarda, "Industrial Restructuring and the Changing Location of Jobs." The figures are in 1989 dollars.

61. Grant and Wallace, "The Political Economy of Manufacturing Growth and Decline."

62. Freeman, "How Labor Fares in Advanced Economies."

63. Crystal, *In Search of Excess*.

64. U.S. Bureau of Census, *Statistical Abstract 1994*, table 1370.

65. U.S. Bureau of Labor Statistics, "New Productivity Data," tables 3–6.

66. Technically these figures are not strict cohort measures. Rather they were arrived at by dividing total enrollments for a given year by an estimate of the population in the appropriate age group.

67. Hout, "The Politics of Mobility," p. 10.

68. Hout, "Expanding Universalism, Less Structural Mobility."

69. Mare, "Changes in Educational Attainment and School Enrollment"; Hout et al., "Making the Grade"; Hout, "Expanding Universalism, Less Structural Mobility."

70. See Lucas, "Educational Transitions of 1980 Sophomores."

71. Levy, *Dollars and Dreams*, p. 123.

72. Katz et al., "A Comparison of Changes in the Structure of Wages in Four OECD Countries," argue that wage differentials between more and less educated workers declined in the United States, Britain, Japan, and France in the 1970s. In the 1980s, when the relative supply of educated workers fell behind demand, educational wage differentials increased sharply in the United States and Britain and modestly in Japan. But inequality was held in check in France because of increases in the national minimum wage and the strength of French labor unions.

73. Levy, "Incomes and Income Inequality."

74. Ibid.

75. Müller and Karle, "Social Selection in Educational Systems in Europe."

76. Arum and Shavit, "Secondary Vocational Education."

77. Will, "What's Behind Income Disparity."

78. On this general point about markets, see Granovetter, "Economic Action and Social Structure."

CHAPTER 7

1. E.g., Coleman and Hoffer, *Public and Private High Schools.*

2. Bryk and Raudenbush, "Toward a More Appropriate Conceptualization of Research on School Effects."

3. Heyns, *Summer Learning and the Effects of Schooling* and "Schooling and Cognitive Development"; Entwistle and Alexander, "Summer Setback" and "Winter Setback."

4. *The Bell Curve,* p. 437.

5. See Alexander and Cook, "Curricula and Coursework"; Gamoran and Mare, "Secondary School Tracking and Educational Inequality"; Garet and DeLany, "Students, Courses, and Stratification"; Hauser, Sewell, and Alwin, "High School Effects on Achievement"; Oakes, *Keeping Track*; Rosenbaum, "Track Misperceptions and Frustrated College Plans"; and Lucas and Gamoran, "Race and Track Assignment."

6. Higher class includes students who were two standard deviations above the middle class, while lower class represents students whose families were two standard deviations below the middle class. The source is Gamoran and Mare, "Secondary School Tracking and Educational Inequality," tables 1 and 3, with table 3 probit coefficients transformed into logit coefficients.

7. Alexander, Cook, and McDill, "Curriculum Tracking and Educational Stratification"; Alexander and Cook, "Curricula and Coursework"; Gamoran, "The Stratification of High School Learning Opportunities"; and Kerckhoff, "Institutional Arrangements and Stratification Processes."

8. E.g., Gamoran and Mare, "Secondary School Tracking and Educational Inequality."

9. E.g., Barr and Dreeben, *How Schools Work*; Gamoran, "Instructional and Institutional Effects of Ability-Grouping"; Weinstein, "Reading Group Membership in First Grade."

10. See Kerckhoff, "Institutional Arrangements and Stratification Processes."

11. For overviews, see Schaie, "Intellectual Development in Adulthood"; and Woodruff-Pak, "Aging and Intelligence." For an example, see Cornelius and Caspi, "Everyday Problem Solving in Adulthood and Old Age." On five hours of training, see Schaie, "The Course of Adult Intellectual Development," p. 311.

12. Schaie, "Intellectual Development in Adulthood," p. 300; Schooler, "Psychosocial Factors and Effective Cognitive Functioning in Adulthood," pp. 351–55; and Schaie, "Midlife Influences upon Intellectual Functioning in Old Age."

13. The quote is from Schooler, "Psychosocial Factors and Effective Cognitive Functioning in Adulthood," p. 349, which summarizes much of the research to about 1990. (The ellipses in the quotation are for the word "men's." Since 1990, the research has been replicated among women as well.) More recent studies include Kohn et al., "Position in the Class Structure and Psychological Functioning in the United States, Japan, and Poland"; and Naoi and Schooler, "Psychological Consequences of Occupational Conditions among Japanese Wives." Earlier, important

studies in this project were Kohn and Schooler, "Occupational Experience and Psychological Functioning" and "The Reciprocal Effects of the Substantive Complexity of Work and Intellectual Flexibility."

"Intellectual flexibility" is basically a synonym for intelligence. The measure typically combines four components: a brief, nonverbal intelligence test, such as the Embedded Figures test; a coding of how well interviewees did in answering a problem-solving question (such as asking respondents to discuss the pros and cons of permitting cigarette advertising); a score for how often interviewees answered "agree" to questions—a negative indicator of their thoughtfulness; and the interviewers' rating of the interviewees' intelligence.

14. Kohn, *Class and Conformity*, p. 190.

15. Ibid., p. 200.

16. Humphreys, "Intelligence," p. 200ff.

17. Psychologists Jerome and Dorothy Singer report that preschool television-watching seems to improve the reading comprehension of poorer children but impair that of middle-class students ("Psychologists Look at Television").

CHAPTER 8

1. We recognize the controversy about group labels such as "black," "African American," and "Latino American." That very controversy testifies to how race is a socially constructed category. We will use black and African American interchangeably here. We prefer "Latino American" in order to note that we are referring to the Western Hemisphere peoples, not all Spanish-speakers. Even then, we focus mostly on Mexican Americans.

2. Lee, "Koreans in Japan and the United States"; DeVos and Wetheral, *Japan's Minorities*; and Rohlen, "Education, Policies, and Prospects."

3. On test scores: Sowell, "New Light on Black I.Q."; on school performance, Lieberson, *A Piece of the Pie*, table 8.13.

4. For example, Brigham, *A Study in American Intelligence*; see also Kamin, "The Pioneers of IQ Testing."

5. We draw here, in part, on the work of John Ogbu, of which more later.

6. There are many library shelves full of books on the history of African Americans. A sample of helpful readings includes Fredrickson, *White Supremacy*; Genovese, *Roll, Jordan, Roll*; Dollard, *Caste and Class in a Southern Town*; and Litwack, *Been in the Storm So Long*.

7. See discussion in Jankowski, "The Rising Significance of Status in U.S. Race Relations."

8. Ogbu, *Minority Education and Caste*; Gibson and Ogbu (eds.), *Minority Status and Schooling*.

9. On Mexican Americans, see David Montejano, *Anglos and Mexicans in the Making of Texas*; Acuna, *Occupied America*; Barrera, *Race and Class in the Southwest*; and McWilliams, *Factories in the Field*.

10. See Montejano, *Anglos and Mexicans*, p. 79.

11. Although Anglo Americans thought Mexicans were a hybrid race that was inferior to theirs, several historical idiosyncrasies allowed them to postpone the aggressive oppression and the concomitant racial justifications until the 1920s. One was the fact that there was a Mexican upper class that was well established when the Anglos conquered the Southwest. Second, Anglo ranchers befriended (even married into, when it was expedient) the Mexican upper class—whom they preferred to call Spanish, i.e., European—and adopted the same feudal system that had been in place before their arrival. Thus the socioeconomic conditions of the Southwest permitted the white population to adopt a less brutal strategy than existed in the South, at least at first.

12. Indeed, there is evidence that freed blacks in the North were advancing quite well, despite severe discrimination, during the nineteenth century; see Lieberson, *A Piece of the Pie*.

13. On the Mexican case, see, for example, Gonzales, "Racial Intelligence Testing and the Mexican People."

14. See Montejano, *Anglos and Mexicans*, pp. 157–256, 262–87.

15. Lieberson, *A Piece of the Pie*, pp. 137–51.

16. For African Americans, see Van Woodward, *The Strange Career of Jim Crow*; and for Mexicans, see Taylor, *Mexican Labor in the United States*.

17. We borrow the phrase from Massey and Denton's major book.

18. Lieberson, *A Piece of the Pie*, pp. 170–72. Also, as late as 1970, northern-educated blacks were far ahead of southern-educated blacks (Lieberson, "Generational Differences").

19. See Kardiner and Ovesey, *The Mark of Oppression*; Clark, *Dark Ghetto*.

20. Card and Krueger, "Trends in Relative Black-White Earnings Revisited"; Smith and Welch, "Black Economic Progress after Myrdal"; Donahue and Heckman, "Continuous versus Episodic Change."

21. The literature is well summarized in Massey and Denton, *American Apartheid*, pp. 96–109. Other studies show discrimination statistically, such as Squires and Velez, "Insurance Redlining in the Transformation of the Urban Metropolis."

22. In addition to Massey and Denton, *American Apartheid*, see, for example, Leonard, "The Interaction of Residential Segregation and Employment Discrimination"; Massey et al., "The Effect of Residential Segregation on Social and Economic Well-Being"; Alba and Logan, "Minority Proximity to Whites in Suburbs"; Logan and Alba, "Locational Returns to Human Capital"; Rosenbaum, "The Constraints on Minority Housing Choices"; Massey et. al, "Migration, Segregation, and the Geographic Concentration of Poverty"; and Amin and Mariam, "Racial Differences in Housing."

23. See Bloch, *Antidiscrimination Law and Minority Employment,* chap. 3.

24. Kirschenman and Neckerman, " 'We'd Love to Hire Them, But . . .' "; see also Newman and Lennon, "The Job Ghetto"; Lucas, "Effects of Race and Gender Discrimination."

25. Newman and Lennon, "The Job Ghetto."

26. Uchitelle, "Union Goal of Equality Fails the Test of Time."

27. Kahn, "Discrimination in Baseball."

28. On the decline in prejudice, see Schuman et al., *Racial Attitudes in America*. Still, by one estimate, about 12 percent of American whites are "extremely prejudiced" (Sniderman and Piazza, *The Scar of Race*).

29. Crosby et al., "Recent Unobtrusive Studies."

30. Feagin, "The Continuing Significance of Race."

31. The source for these data is Massey and Denton, *American Apartheid*, pp. 24, 48, 64, and Harrison and Weinberg, "Racial and Ethnic Residential Segregation." These scholars draw on the foundational work of Lieberson, *A Piece of the Pie*.

32. Lieberson, *Ethnic Patterns*.

33. See, e.g., Jargowsky, "Ghetto Poverty among Blacks in the 1980s."

34. Massey and Denton, *American Apartheid*.

35. See Kozol, *Savage Inequalities*; and Orfield and Ashkinaze, *The Closing Door*.

36. Ogbu, *Minority Education and Caste*; Gibson and Ogbu (eds.), *Minority Status and Schooling*; Ogbu, "Minority Status." It is odd that Herrnstein and Murray describe Ogbu's theory in one paragraph (p. 307) and then simply go on. It is odd, because Ogbu provides a complete alternative explanation to their own for the racial differences in achievement scores. The authors of *The Bell Curve* note that explanation but never deal with it.

37. Quoted by the Associated Press in "Powell's Book Criticizes Bush for Horton Ad," *San Francisco Chronicle*, September 15, 1995, p. 2.

38. See Duncan, "Inheritance of Poverty or Inheritance of Race?"; Farley, *Black and White Income*; Borjas and Tienda, *Hispanics in the U.S. Economy*; Tienda et al., "Schooling, Color, and the Labor Force Activity of Women"; Bianchi, "Changing Economic Roles of Women and Men"; Blackburn et al., "Changes in Earnings Differentials in the 1980s"; and Bound and Johnson, "Changes in the Structure of Wages during the 1980s."

39. For comprehensive views of racial differences in America today, see Hacker, *Two Nations*, and Farley and Allen, *The Color Line and the Quality of Life in America*.

40. See the literature on the "underclass"—for example, Jencks and Peterson (eds.), *The Urban Underclass*; Wilson, *The Truly Disadvantaged*; and Wilson (ed.), *The Ghetto Underclass*. For some of the latest data, see Jargowsky, "Ghetto Poverty."

41. Nisbett, "Race, IQ, and Scientism," esp. pp. 48–53; Hauser, Review of *The Bell Curve*.

42. Quoted in Wolf, *Urban Village*, p. 138.

43. Quoted by Higham, *Strangers in the Land*, p. 143.

44. Sowell, "Ethnicity and IQ," pp. 32–36; also Sowell, "New Light on Black I.Q."; and Lieberson, *A Piece of the Pie*.

45. Brigham, in *The Bell Curve Debate*; Pearson quoted by Barkan, *The Retreat of Scientific Racism*, p. 156.

46. Ogbu, *Minority Education*; Gibson and Ogbu (eds.), *Minority Status and Schooling*.

47. Although the image may exist that these two groups of Jews differ physically, in fact their historical origins belie any major genetic differences.

48. See, in particular, Dar and Resh, "Socioeconomic and Ethnic Gaps."

49. DeVos and Wetherall, *Japan's Minorities*; Lee and DeVos, *Koreans in Japan*.

50. DeVos and Wetherall, *Japan's Minorities*; Shimahara, "Social Mobility and Education"; Kristof, "Japanese Outcasts."

51. Verster and Prinsloo, "The Diminishing Test Performance Gap."

52. This explanation also helps us understand why, around the world, urban children almost always score higher than rural children. Urban residents are usually privileged over rural ones. (It may also help explain why American children from the South score lower than children from other regions—Parcel and Geschwender, "Explaining Southern Disadvantage.")

53. Research shows that African Americans, for example, have historically placed a high value on attaining an education. See, e.g., Lieberson, *Piece of the Pie*.

54. Oliver and Shapiro, *Black Wealth/White Wealth*, table 5.2. The white couple has liquid assets of $1,150, the black couple, $0.

55. See Storfer, *Intelligence and Giftedness*, for a discussion of some of these variables.

56. Crane, "Race and Children's Cognitive Test Scores." See also Brooks-Gunn, "Ethnic Differences in Children's Intelligence Test Scores," and Crane, "Exploding the Myth of Scientific Support."

57. Crane, "Race and Children's Cognitive Test Scores," for example, statistically controls for the contribution that mothers' test scores make to the environmental variables.

58. See, for example, Massey and Denton, *American Apartheid*; Massey and Gross, "Explaining Trends in Racial Segregation"; Massey et al., "Migration, Segregation, and the Geographic Concentration of Poverty"; and Farley and Frey, "Changes in the Segregation of Whites from Blacks during the 1980s."

59. See, for example, Massey and Fong, "Segregation and Neighborhood Quality"; Alba and Logan, "Minority Proximity to Whites in Suburbs"; Massey et al., "The Effect of Residential Segregation on Social and Economic Well-Being"; Logan and Alba, "Locational Returns to Human Capital"; and Alba et al., "Living with Crime."

60. Jargowsky, "Ghetto Poverty"; Massey and Denton, *American Apartheid*.

61. Hoffer et al., "Achievement Growth in Public and Catholic Schools."

62. Boozer and Rouse, "Interschool Variation"; Kozol, *Savage Inequalities*.

63. For example, Anderson, "Neighborhood Effects on Teenage Pregnancy"; Crane, "The Epidemic Theory of Ghettos"; Furstenberg et al., "Race Differences in

the Timing of Adolescent Intercourse"; Brewster, "Race Differences in Sexual Activity among Adolescent Women"; but not Ku et al., "Neighborhood, Family, and Work." See, especially on academic achievement, Quane and Rankin, "Does Living in a Poor Neighborhood?" For overview, see Massey and Denton, *American Apartheid*, ch. 6.

64. The same studies also show that in situations where the stereotype was not salient, black students scored as well as white students, after adjusting for entering SAT scores. See Steele and Aronson, "Contending with a Stereotype"; also Steele, "Race and the Schooling of Black Americans" and "Protective Dis-Identification and Academic Performance." And R. Morin, "Stereotype Straightjackets," *Washington Post*, January 28, 1996, p. C5.

65. See, for example, Fordham and Ogbu, "Black Students' School Success"; Ogbu, "Racial Stratification and Education in the United States" and "Community Forces and Minority Educational Strategies"; Mickelson, "The Attitude-Achievement Paradox among Black Adolescents."

66. The proportion of black and Latino NLSY respondents who scored below chance well exceeded that of whites. About 9 percent of Latino and 12 percent of black test takers scored below chance, and in the last section of the test more than 30 percent of both groups scored below chance. Some of these low-scorers may have been exceedingly "dull," but we suspect that most were exceedingly uninterested in the task, despite the $50 fee—which they received no matter how well they scored. (Bock and Moore, *Advantage and Disadvantage*, p. 95, raise a similar point when they evaluate the race differences on the "Coding Speed" task.)

67. Lee, "Koreans in Japan and the United States"; DeVos and Chung, "Community Life in a Korean Ghetto," p. 247.

68. Willis, *Learning to Labour*.

69. Yossi Shavit, University of Haifa, personal communication.

70. See review of *The Bell Curve* evidence on Asian IQ in Lane, "Tainted Sources."

71. Sowell, "Ethnicity and IQ."

72. Stevenson et al., "Mathematics Achievement of Chinese, Japanese, and American Children"; Schneider and Lee, "A Model for Academic Success"; and Ogbu, "Minority Status," p. 374.

CHAPTER 9

1. Bane and Jencks, "Five Myths," pp. 28, 32.

2. Ibid., p. 40.

3. See, for example, Schwarz, *America's Hidden Success*; Schorr and Schorr, *Within Our Reach*; Mayer and Jencks, "War on Poverty."

4. For sources on Americans' beliefs about inequality, see, for example, Swidler, "Inequality and American Culture"; Hochschild, *What's Fair* and *Facing Up to the American Dream*; Huston, "The American Revolutionaries"; Weiss, *The American*

Myth of Success; Gans, *Middle American Individualism*; Kelley and Evans, "The Legitimation of Inequality"; Rainwater, *What Money Buys*; Coleman and Rainwater, *Social Standing in America*; and Verba and Orren, *Equality in America*.

5. See, especially, Huston, "The American Revolutionaries."

6. There are library shelves full of books on individualism in America. See, for example, Bellah et al., *Habits of the Heart*; Gans, *Middle American Individualism*; Arieli, *Individualism and Nationalism in American Ideology*; Curry and Goodheart (eds.), *American Chameleon*; Curti, *Human Nature in American Thought*; and Triandis, "Cross-Cultural Studies of Individualism and Collectivism."

7. The cases are drawn from a representative national survey conducted by Erik Wright, Michael Hout, and Martín Sánchez Jankowski. We began with men born between 1935 and 1939 who had working-class fathers and some education and selected the one with the median income. He became "John Smith, Jr." Then, based on what we knew about his father and his son, we selected matches for those two out of the sample. Some details are simplified for presentation, but the basic accounts are factual.

8. Children of A's, unless determined to be unfit, would be automatically promoted from B to A at their majority; children of C's could pass a test to get into B and then spend twenty years there on probation before applying to become A's.

9. See Tucker, *The Science and Politics of Racial Research*, pp. 104–5, for McDougall and Cutten; *passim* for other policy discussions.

10. In "Race, Genes, and I.Q.," Murray suggests that blacks and other low-IQ ethnic groups console themselves by celebrating the nonintellectual traits they do have—presumably, for blacks, meaning rhythm and slam-dunking.

11. Illustrative of the research are studies linking economic inequality to rates of violent crime—e.g., Balkwell, "Ethnic Inequality and the Rate of Homicide"; Harper and Steffensmeier, "The Differing Effects of Economic Inequality"; and Hagan and Peterson, *Crime and Inequality*.

12. Granovetter, "Economic Action and Social Structure."

13. See, for example, Rainwater, *What Money Buys*; and Hochschild, *What's Fair?*.

❖ References ❖

Abrams, Charles. 1966. "The Housing Problem and the Negro." *Daedalus* 95:64–76.

Achenbaum, W. A. 1993. "Old Age," pp. 2051–62 in Mary Kupiec Cayton, Elliott J. Gorn, and Peter W. Williams (eds.), *Encyclopedia of American Social History*. New York: Charles Scribner's Sons.

Acuña, Rodolfo. 1988. *Occupied America: A History of Chicanos*. 3d ed. New York: Harper & Row.

Adamovic, Karol. 1979. "Intellectual Development and Level of Knowledge in Gypsy Pupils in Relation to the Type of Education." *Psychologia a Patopsychologia Dietata* 14, 2:169–76 (translated abstract).

Alba, Richard D., and John R. Logan. 1993. "Minority Proximity to Whites in Suburbs: An Individual-Level Analysis of Segregation." *American Journal of Sociology* 98 (May): 1388–1427.

Alba, Richard D., John R. Logan, and Paul E. Bellair. 1994. "Living with Crime: The Implications of Racial/Ethnic Differences in Suburban Location." *Social Forces* 73 (December): 395–434.

Alexander, Karl L., and M. A. Cook. 1982. "Curricula and Coursework: A Surprise Ending to a Familiar Story." *American Sociological Review* 47:626–40.

Alexander, Karl L., M. A. Cook, and E. L. McDill. 1978. "Curriculum Tracking and Educational Stratification." *American Sociological Review* 43:47–66.

Althschuler, Glenn C., and Jan M. Saltzgaber. 1988. "The Limits of Responsibility: Social Welfare and Local Government in Seneca County, New York, 1860–1875." *Journal of Social History* 22 (Spring 1988): 515–38.

Alwin, Duane. 1991. "Family of Origin and Cohort Differences in Verbal Ability." *American Sociological Review* 56:625–38.

Amin, Ruhul, and A. G. Mariam. 1987. "Racial Differences in Housing: An Analysis of Trends and Differentials." *Urban Affairs Quarterly* 22 (March): 363–76.

Anderson, Elijah. 1978. *A Place on the Corner*. Chicago: University of Chicago Press.

———. 1990. *Streetwise: Race, Class, and Change in an Urban Community*. Chicago: University of Chicago Press, 1990.

———. 1991. "Neighborhood Effects on Teenage Pregnancy," pp. 375–98 in Christopher Jencks and Paul E. Peterson (eds.), *The Urban Underclass*. Washington, D.C.: Brookings Institution.

Anyon, Jean. 1981. "Social Class and School Knowledge." *Curriculum Inquiry* 11,1:3–42.

Arieli, Yehoshua. 1964. *Individualism and Nationalism in American Ideology*. Cambridge: Harvard University Press.

277

Arum, Richard, and Yossi Shavit. 1995. "Secondary Vocational Education and the Transition from School to Work." *Sociology of Education* 68:187–204.

Arvey, Richard D., et al., "Mainstream Science on Intelligence," letter to *Wall Street Journal*, December 13, 1994.

Atkinson, Anthony B., Lee Rainwater, and Timothy M. Smeeding. 1995. "Income Distribution in Advanced Economies: The Evidence from the Luxembourg Study." Working Paper no. 120, Luxembourg Income Study, Luxembourg.

Baker, David P., and Deborah Perkins Jones. 1993. "Creating Gender Equality: Cross-National Gender Stratification and Mathematical Performance." *Sociology of Education* 66 (April): 91–103.

Baker, Dean. 1996. "The Inflated Case against the CPI." *The American Prospect*, no. 24 (Winter): 86–89.

Balkwell, James W. 1990. "Ethnic Inequality and the Rate of Homicide." *Social Forces* 69 (September): 53–70.

Bane, Mary Jo, and David B. Ellwood. 1994. *Welfare Reform*. New York: Russell Sage Foundation.

Bane, Mary Jo, and Christopher Jencks. 1973. "Five Myths about Your IQ." *Harper's* (February): 28–40.

Barkan, Elazar. 1992. *The Retreat of Scientific Racism*. New York: Cambridge University Press.

Barr, Rebecca, and Robert Dreeben. 1983. *How Schools Work*. Chicago: University of Chicago Press.

Barrera, Mario. 1979. *Race and Class in the Southwest: A Theory of Racial Inequality*. South Bend: University of Notre Dame Press.

Bellah, Robert, Richard Madsen, William Sullivan, Ann Swidler, and Steven Tipton. 1985. *Habits of the Heart: Individualism and Commitment in American Life*. Berkeley: University of California Press.

Benson, Ciarán. 1995 (1987). "Ireland's 'Low' IQ: A Critique," pp. 222–33 in Russell Jacoby and Naomi Glauberman (eds.), *The Bell Curve Debate*. New York: Times Books.

Bernstein, Basil B. 1977. *Class, Codes, and Control*. 2d rev. ed. London: Routledge and Kegan Paul.

Bianchi, Suzzanne. 1995. "Changing Economic Roles of Women and Men," pp. 107–54 in Reynolds Farley (ed.), *State of the Union*. New York: Russell Sage Foundation.

Blackburn, McKinley L., David E. Bloom, and Richard B. Freeman. 1991. "Changes in Earnings Differentials in the 1980s: Concordance, Convergence, Causes, and Consequences." National Bureau of Economic Research Working Paper. Cambridge, Mass.: National Bureau of Economic Research.

———. 1992. "Changes in Earnings Differentials in the 1980s." National Bureau of Economic Research Working Paper no. 3901, November.

Blank, Rebecca. 1994. "Does a Larger Social Safety Net Mean Less Economic

Flexibility?" pp. 157–87 in Richard Freeman (ed.), *Working under Different Rules*. New York: Russell Sage Foundation.

Bloch, Farrell. 1994. *Antidiscrimination Law and Minority Employment*. Chicago: University of Chicago Press.

Bluestone, Barry. 1995. "The Inequality Express." *The American Prospect* (Winter): 81–93.

Blumin, Stuart M. 1989. *The Emergence of the Middle Class: Social Experience in the American City, 1760–1900*. New York: Cambridge University Press.

Bock, Kenneth. 1994. *Human Nature Mythology*. Urbana: University of Illinois Press.

Bock, R. Darrel, and Elsie G. J. Moore. 1986. *Advantage and Disadvantage: A Profile of American Youth*. Hillsdale, N.J.: Lawrence Erlbaum Associates.

Borjas, George, and Marta Tienda, 1986. *Hispanics in the U.S. Economy*. Orlando: Academic Press.

Boozer, Michael, and Cecilia Rouse. 1995. "Interschool Variation in Class Size." NBER Working Paper No. 5144. Cambridge, Mass., National Bureau of Economic Research.

Bound, John, and George Johnson. 1992. "Changes in the Structure of Wages during the 1980s." *American Economic Review* 82:371–92.

Bradsher, Keith. 1995. "America's Opportunity Gap," *New York Times*, June 4: E-4.

Brewster, Karin L. 1994. "Race Differences in Sexual Activity among Adolescent Women: The Role of Neighborhood Characteristics." *American Sociological Review* 59 (June): 408–24.

Brigham, Carl C. 1995 (1923). *A Study of American Intelligence*, excerpted, pp. 571–82, in Russell Jacoby and Naomi Glauberman (eds.), *The Bell Curve Debate*. New York: Times Books.

Brock, William R. 1988. *Welfare, Democracy, and the New Deal*. New York: Cambridge University Press.

Brooks-Gunn, Jeanne. 1995. "Ethnic Differences in Children's Intelligence Test Scores: The Role of Economic Deprivation, Home Environment and Maternal Characteristics." Paper presented to Seminar on Meritocracy and Equality, University of Chicago, May 26.

Brooks-Gunn, Jeanne, Greg J. Duncan, Pamela K. Klebanov, and Naomi Sealand. 1993. "Do Neighborhoods Influence Child and Adolescent Development?" *American Journal of Sociology* 99:353–95.

Browne, Irene. 1995. "The Baby Boom and Trends in Poverty, 1967–1987." *Social Forces* 73 (March): 1071–95.

Bruchey, Stuart. 1990. *Enterprise: The Dynamic Economy of a Free People*. Cambridge: Harvard University Press.

Bryk, Anthony S., and Stephan W. Raudenbush. 1988. "Toward a More Appropriate Conceptualization of Research on School Effects: A Three-Level Hierarchical Linear Model." *American Journal of Education* 97:65–108.

Bumpass, Larry. 1990. "What Is Happening to the Family? Interactions between Demographic and Institutional Change," *Demography* 27:483–98.

Burtless, Gary. 1994. "Public Spending on the Poor: Historical Trends and Economic Limits," pp. 51–84 in Sheldon H. Danziger, Gary D. Sandefur, and Daniel H. Weinberg (eds.), *Confronting Poverty: Prescriptions for Change*. Cambridge: Harvard University Press.

Card, David. 1992. "The Effect of Unions on the Distribution of Wages." National Bureau of Economic Research Working Paper no. 4195, October.

Card, David, and Richard B. Freeman. 1994. "Small Differences That Matter: Canada vs. the United States," pp. 180–222 in Richard Freeman (ed.), *Working under Different Rules*. New York: Russell Sage Foundation.

Card, David, and Alan B. Krueger. 1993. "Trends in Relative Black-White Earnings Revisited." *American Economic Review* 83 (May): 85–91.

Ceci, Stephen J. 1990. *On Intelligence . . . More or Less: A Bio-ecological Treatise on Intellectual Development*. Englewood Cliffs, N.J.: Prentice-Hall.

Ceci, Stephen J., and Anotonio Roazzi. 1994. "The Effects of Context on Cognition: Postcards from Brazil," pp. 74–104 in Robert J. Sternberg and Richard K. Wagner (eds.), *Mind in Context: Interactionist Perspectives on Human Intelligence*. New York: Cambridge University Press.

Chafe, William. 1986. *The Unfinished Journey: America since World War II*. New York: Oxford University Press.

Chambers, Clarke A. 1963. *Seedtime of Reform*. Minneapolis: University of Minnesota Press.

Chang, Roberto. 1994. "Income Inequality and Economic Growth: Evidence and Recent Theories." *Economic Review* (Federal Reserve Bank of Atlanta) 79 (July/August): 1–10.

Chapin, Theodore. 1989. "The Relationship of Trait Anxiety and Academic Performance to Achievement Anxiety." *Journal of College Student Development* 30:229–36.

Chevan, Albert. 1989. "The Growth of Home Ownership: 1940–1980." *Demography* 26 (May): 249–66

Church, Avery G. 1976. "Academic Achievement, IQ, Level of Occupational Plans, and Ethnic Stereotypes for Anglos and Navahos in a Multi-ethnic High School." *Southern Journal of Educational Research* (Summer): 184–201.

Clark, Kenneth B. 1965. *Dark Ghetto: Dilemmas of Social Power*. New York: Harper & Row.

Clark, Lesley A., and Graeme S. Halford. 1983. "Does Cognitive Style Account for Cultural Differences in Scholastic Achievement?" *Journal of Cross-Cultural Psychology* 14 (September): 279–96.

Clarke, George R. G. 1995. "More Evidence on Income Distribution and Growth." *Journal of Development Economics* 47:403–27.

Cohen, Patricia Cline. 1982. *A Calculating People: The Spread of Numeracy in Early America*. Chicago: University of Chicago Press.

Coleman, James S., and Thomas Hoffer. 1987. *Public and Private High Schools: The Impact of Communities.* New York: Basic Books.

Coleman, Richard Patrick, and Lee Rainwater. 1978. *Social Standing in America: New Dimensions of Class.* New York: Basic Books.

Cookson, Peter W., and Caroline Hodges Persell. 1985. *Preparing for Power.* New York: Basic Books.

Coontz, Stephanie. 1992. *The Way We Never Were: American Families and the Nostalgia Trap.* New York: Basic Books.

Cornelius, Steven W., and Avshalom Caspi. 1987. "Everyday Problem Solving in Adulthood and Old Age." *Psychology and Aging* 2 (June): 144–53.

Cowherd, Douglas M., and David I. Levine. 1992. "Product Quality and Pay Equity." *Administrative Science Quarterly* 37 (June): 302–30.

Crane, Jonathan. 1991. "The Epidemic Theory of Ghettos and Neighborhood Effects of Dropping Out and Teenage Childbearing." *American Journal of Sociology* 96 (March): 1126–59.

———. 1994. "Exploding the Myth of Scientific Support for the Theory of Black Intellectual Inferiority." *Journal of Black Psychology* 20 (May): 189–209.

———. 1995. "Race and Children's Cognitive Test Scores: Empirical Evidence that Environment Explains the Entire Gap." Paper presented to Seminar on Meritocracy and Equality, University of Chicago, May 26.

Crosby, Faye, Stephanie Bronmley, and Leonard Saxe. 1980. "Recent Unobtrusive Studies of Black and White Discrimination and Prejudice: A Literature Review." *Psychological Bulletin* 87, 3:546–63.

Crystal, Graef A. 1991. *In Search of Excess: The Overcompensation of American Executives.* New York: W. W. Norton.

Curry, Richard O., and Lawrence B. Goodheart (eds.). 1991. *American Chameleon: Individualism in Trans-National Context.* Kent, Ohio: Kent State University Press.

Curti, Merle. 1980. *Human Nature in American Thought: A History.* Madison: University of Wisconsin Press.

Cutler, David M., and Lawrence F. Katz. 1992. "Rising Inequality? Changes in the Distribution of Income and Consumption in the 1980s." *American Economic Review* 82 (May): 546–51.

Danziger, Sheldon, and Peter Gottschalk. 1993. "Introduction," pp. 6–9 in Sheldon Danziger and Peter Gottschalk (eds.), *Uneven Tides: Rising Inequality in America.* New York: Russell Sage Foundation.

Danziger, Sheldon, and Peter Gottschalk (eds.). 1993. *Uneven Tides: Rising Inequality in America.* New York: Russell Sage Foundation.

Dar, Yehezekal, and Nura Resh. 1991. "Socioeconomic and Ethnic Gaps in Academic Achievement in Israeli Junior High Schools," pp. 322–37 in Nico Bleichrodt and Peter Drenth (eds.), *Contemporary Issues in Cross-cultural Psychology.* Amsterdam: Swets & Zeitlinger.

Das, J. P., and Amulya Kanti Satpathy Khurana. 1988. "Caste and Cognitive Processes," pp. 487–508 in S. H. Irvine and J. W. Berry (eds.), *Human Abilities in Cultural Context*. Cambridge: Cambridge University Press.

Das, Sohani. 1994. "Level-I Abilities of Socially Disadvantaged Children: Effects of Home-Environment, Caste and Age." *Social Science International* 10, 1–2:69–74 (abstract).

Das, Sohani, and Brahmananda Padhee. 1993. "Level II Abilities of Socially Disadvantaged Children: Effects of Home-Environment, Caste and Age." *Journal of Indian Psychology* 11, 1–2:38–43 (abstract).

Davis, James A., and Tom W. Smith. 1994. *General Social Survey Cumulative Codebook* [MRDF]. Storrs, Conn.: Roper.

DeVos, George, and Daekyun Chung. 1981. "Community Life in a Korean Ghetto," pp. 223–51 in Changsoo Lee and George DeVos (eds.), *Koreans in Japan: Ethnic Conflict and Accommodation*. Berkeley: University of California Press.

DeVos, George A., and William O. Wetherall. 1983. *Japan's Minorities*. London: Minority Rights Group.

Dickens, William T., Thomas J. Kane, and Charles L. Schultze. 1995. "Does *The Bell Curve* Ring True?" *The Brookings Review* (Summer): 19–22.

———. 1996. *Does The Bell Curve Ring True?* Washington, D.C.: Brookings Institution.

DiPrete, Thomas A., and David B. Grusky. 1990. "Structure and Trend in the Process of Social Stratification." *American Journal of Sociology* 96 (July): 107–44.

Dollard, John. 1988 (1937). *Caste and Class in a Southern Town*. Madison: University of Wisconsin Press.

Donahue, John J., III, and James Heckman. 1994. "Continuous versus Episodic Change: The Impact of Civil Rights Policy on the Economic Status of Blacks," pp. 183–206 in Paul Burstein (ed.), *Equal Employment Opportunity: Labor Market Discrimination and Public Policy*. New York: Aldine de Gruyter.

Downey, Douglas B. 1995. "When Bigger is Not Better: Family Size, Parental Resources, and Children's Educational Performance." *American Sociological Review* 60 (October): 746–61.

Dreier, Peter, and John Atlas. 1995. "Housing Policy's Moment of Truth." *The American Prospect* 22 (Summer): 68–77.

Dreze, Jean, and Amartya Sen. 1989. *Hunger and Public Action*. Oxford: Clarendon Press.

Duncan, Greg J., and Willard Rodgers. 1991. "Has Children's Poverty Become More Persistent?" *American Sociological Review* 56 (July): 538–50.

Duncan, Greg J., and Ken R. Smith. 1989. "The Rising Affluence of the Elderly: How Far, How Fair, How Frail?" *Annual Review of Sociology* 15: 261–89.

Duncan, Otis Dudley. 1969. "Inheritance of Poverty or Inheritance of Race?" pp. 85–110 in Daniel Patrick Moynihan (ed.), *On Understanding Poverty*. New York: Basic Books.

Duncan, Otis Dudley, David L. Featherman, and Beverly Duncan. 1972. *Socioeconomic Background and Achievement*. New York: Academic Press.

Duster, Troy. 1995. Review of *The Bell Curve. Contemporary Sociology* 24 (March): 158–61.

———. 1995. "The Advantages of White Males." *San Francisco Chronicle*, January 19.

Educational Testing Service. 1995. *Admission Officer's Handbook for the SAT Program*. Princeton: Educational Testing Service.

Entwistle, Doris R., and Karl L. Alexander. 1992. "Summer Setback: Race, Poverty, School Composition, and Mathematics Achievement in the First Two Years of School." *American Sociological Review* 57:72–84.

———. 1994. "Winter Setback: The Racial Composition of Schools and Learning to Read." *American Sociological Review* 59 (June): 446–60.

Fantasia, Rick. 1988. *Cultures of Solidarity: Consciousness, Action, and Contemporary American Workers*. Berkeley: University of California Press.

Farley, Reynolds. 1982. *Black and White Income*. Cambridge: Harvard University Press.

Farley, Reynolds, and Richard R. Allen. 1987. *The Color Line and the Quality of Life in America*. Cambridge: Harvard University Press.

Farley, Reynolds, and William H. Frey. 1994. "Changes in the Segregation of Whites from Blacks during the 1980s: Small Steps toward a More Integrated Society." *American Sociological Review* 59 (February): 23–45.

Feagin, Joe. 1991. "The Continuing Significance of Race—Antiblack Discrimination in Public Places." *American Sociological Review* 56 (February): 101–16.

Featherman, David L., and Robert M. Hauser. 1978. *Opportunity and Change*. New York: Acadmic Press.

Fiedler, Fred E., and Thomas G. Link. 1994. "Leader Intelligence, Interpersonal Stress, and Task Performance," pp. 152–70 in Robert J. Sternberg and Richard K. Wagner (eds.), *Mind in Context: Interactionist Perspectives on Human Intelligence*. New York: Cambridge University Press.

Flynn, James R. 1987. "Massive IQ Gains in 14 Nations: What IQ Tests Really Measure," *Psychological Bulletin* 101, 2:171–91.

Fogel, Robert William. 1986. "Nutrition and the Decline in Mortality since 1700: Some Preliminary Findings," pp. 439–556 in Stanley L. Engerman and Robert E. Gallman (eds.), *Long-Term Factors in American Economic Growth*. NBER Studies in Income and Wealth, vol. 51. Chicago: University of Chicago Press.

Ford, Ramona. 1988. *Work, Organization, and Power*. Boston: Allyn and Bacon.

Fordham, Signithia, and John U. Ogbu. 1986. "Black Students' School Success: Coping with the 'Burden of "Acting White."'" *The Urban Review* 18:176–206.

Fort, Rodney. 1992. "Pay and Performance," pp. 134–57 in Paul M. Sommers (ed.), *Diamonds are Forever: The Business of Baseball*. Washington, D.C.: Brookings Institution.

Fraser, Steven (ed.). 1995. *The Bell Curve Wars*. New York: Basic Books.

Fredrickson, George M. 1981. *White Supremacy: a Comparative Study in American and South African History*. New York: Oxford University Press.

Freeman, Richard B. 1993. "How Much Has De-unionization Contributed to the Rise in Male Earning Inequality?" pp. 133–63 in Sheldon Danziger and Peter Gottschalk (eds.), *Uneven Tides: Rising Inequality in America.* New York: Russell Sage Foundation.

————. 1994. "How Labor Fares in Advanced Economies," pp. 1–28 in Richard Freeman (ed.), *Working under Different Rules.* New York: Russell Sage Foundation.

Freeman, Richard (ed.) 1994. *Working under Different Rules.* New York: Russell Sage.

Freeman, Richard B., and Lawrence F. Katz. 1994. "Rising Wage Inequality: The United States vs. Other Advanced Countries," pp. 29–62 in Richard Freeman (ed.), *Working under Different Rules.* New York: Russell Sage.

Freeman, Richard B., and James L. Medoff. 1984. *What Do Unions Do?* New York: Basic Books.

Frum, David. 1995. "Righter than Newt." *Atlantic Monthly* (March): 81.

Furstenberg, Frank F., Jr., S. Philip Morgan, Kristin A. Moore, and James L. Peterson. 1987. "Race Differences in the Timing of Adolescent Intercourse," *American Sociological Review* 52 (August): 511–18.

Gamoran, Adam. 1986. "Instructional and Institutional Effects of Ability-Grouping." *Sociology of Education* 59:185–98.

————. 1987. "The Stratification of High School Learning Opportunities." *Sociology of Education* 60:135–55.

Gamoran, Adam, and Robert D. Mare. 1989. "Secondary School Tracking and Educational Inequality: Compensation, Reinforcement, or Neutrality." *American Journal of Sociology* 94:1146–83.

Gans, Herbert. 1988. *Middle American Individualism: The Future of American Individualism.* New York: Free Press.

Gardner, Howard. 1993. *Multiple Intelligences: The Theory in Practice.* New York: Basic Books.

Garet, Michael S., and Brian DeLaney. 1988. "Students, Courses, and Stratification." *Sociology of Education* 61:61–77.

Garfinkel, Irwin, and Sara McLanahan. 1986. *Single Mothers and Their Children.* Washington, D.C.: Urban Institute.

————. 1994. "Single-Mother Families, Economic Insecurity, and Government Policy," pp. 205–25 in Sheldon H. Danziger, Gary D. Sandefur, and Daniel H. Weinberg (eds.), *Confronting Poverty: Prescriptions for Change.* Cambridge: Harvard University Press.

Ganzeboom, Harry B. G., Donald J. Treiman, and Wout C. Ultee. 1991. "Comparative Intergenerational Stratification Research." *Annual Review of Sociology* 17:277–302.

Genovese, Eugene D. 1974. *Roll, Jordan, Roll; the World the Slaves Made.* New York: Pantheon Books.

Gibson, Margaret A., and John U. Ogbu (eds.). 1991. *Minority Status and School-*

ing: A Comparative Study of Immigrants and Involuntary Minorities. New York: Garland.

Goldberger, Arthur S., and Charles F. Manski. 1995. "Review Article: *The Bell Curve* by Herrnstein and Murray." *Journal of Economic Literature* 33 (June): 762–76.

Gonzales, Gilbert C. 1992. "Racial Intelligence Testing and the Mexican People." *Explorations in Ethnic Studies* 5 (July): 36–55.

Gottschalk, Peter, and Mary Joyce. 1995. "The Impact of Technological Change, Deindustrialization, and Internationalization of Trade on Earnings Inequality: An International Perspective," pp. 197–228 in Katherine McFate, Roger Lawson, and William Julius Wilson (eds.), *Poverty, Inequality, and the Future of Social Policy*. New York: Russell Sage Foundation.

Gottschalk, Peter, and Timothy Smeeding. 1995. "Cross-National Comparisons of Levels and Trends in Inequality." Working Paper no. 126, Luxembourg Income Study. Syracuse, N.Y.: Maxwell School of Citizenship and Public Affairs, July.

Gould, Stephen Jay. 1981. *The Mismeasure of Man*. New York: Norton.

Gramlich, Edward M., Richard Kasten, and Frank Sammartino. 1993. "Growing Inequality in the 1980s: The Role of Federal Taxes and Cash Transfers," pp. 225–50 in Sheldon Danziger and Peter Gottschalk (eds.), *Uneven Tides: Rising Inequality in America*. New York: Russell Sage Foundation.

Granovetter, Mark. 1985. "Economic Action and Social Structure: The Problem of Embeddedness." *American Journal of Sociology* 91:481–510.

Grant, Don Sherman, and Michael Wallace. 1994. "The Political Economy of Manufacturing Growth and Decline, 1970–1985." *Social Forces* 73:33–65.

Greven, Philip J., Jr. 1970. *Four Generations: Population, Land, and Family in Colonial Andover, Massachusetts*. Ithaca: Cornell University Press.

Grigsby, William G. 1990. "Housing Finance and Subsidies in the United States," *Urban Studies* 27:831–45.

Griliches, Zvi, and William M. Mason. 1972. "Education, Earnings, and Ability." *Journal of Political Economy* 80:S74–S103.

Gross, Morris B. 1978. "Cultural Concomitants of Preschoolers' Preparation for Learning." *Psychological Reports* 43 (December): 807–13.

Gupta, Anita, and Qamar Jahan. 1989. "Differences in Cognitive Capacity among Tribal and Non-tribal High School Students of Himachal Pradesh." *Manas* 36, 1–2:17–25 (abstract).

Haber, Carole, and Brian Gratton. 1994. *Old Age and the Search for Security*. Bloomington: Indiana University Press.

Hacker, Andrew. 1995. "Caste, Crime, and Precocity," pp. 97–108 in Steven Fraser (ed.), *The Bell Curve Wars*. New York: Basic Books.

———. 1995. *Two Nations: Black and White, Separate, Hostile, Unequal*. Expanded and updated ed. New York: Ballantine Books.

Hagan, John, and Ruth D. Peterson (eds.). 1995. *Crime and Inequality*. Stanford: Stanford University Press.

Hanna, Gila. 1984. "Cross-Cultural Gender Differences in Mathematics Education." *International Journal of Educational Research* 21:417–26.

Hannerz, Ulf. 1969. *Soulside: Inquiries into Ghetto Culture and Community*. New York: Columbia University Press.

Hanson, F. Allan. 1993. *Testing Testing: Social Consequences of the Examined Life*. Berkeley: University of California Press.

Harper, Miles D., and Darrell Steffensmeier. 1992. "The Differing Effects of Economic Inequality and Black and White Rates of Violence." *Social Forces* 70 (June): 1035–54.

Harrison, Roderick J., and Daniel H. Weinberg. 1992. "Racial and Ethnic Residential Segregation in 1990." Paper presented to the Population Association of America.

Hauser, Robert. 1995. Review of *The Bell Curve*. *Contemporary Sociology* 24 (March): 149–53.

Hauser, Robert M., and Wendy Y. Carter. 1995. "*The Bell Curve* as a Study of Social Stratification." Paper presented at the annual meeting of the American Sociological Association, Washington, D.C.

Hauser, Robert M., William Sewell, and Duane Alwin. 1976. "High School Effects on Achievement," pp. 309–41 in William Sewell, Robert M. Hauser, and David Featherman (eds.), *Schooling and Achievement in American Society*. New York: Academic Press.

Hauser, Robert, Donald Treiman, and David Featherman. 1977. "The Intergenerational Transmission of Income Status," pp. 271–301 in Robert Hauser and David Featherman (eds.), *The Process of Stratification*. New York: Academic Press.

Headey, Bruce. 1987. *Housing Policy in the Developed Economy: The United Kingdom, Sweden, and the United States*. New York: St. Martin's Press.

Heckman, James J. 1995. "Lessons from *The Bell Curve*." *Journal of Political Economy* 103 (October): 1091–120.

Heisler, Barbara Schmitter. 1990. "Housing Policy and the Underclass: The United Kingdom, Germany, and the Netherlands." *Journal of Urban Affairs* 16 (3): 203–20.

Herrnstein, Richard J., and Charles Murray. 1994. *The Bell Curve: Intelligence and Class Structure in American Life*. New York: The Free Press.

Heyns, Barbara. 1978. *Summer Learning and the Effects of Schooling*. New York: Academic Press.

———. 1987. "Schooling and Cognitive Development: Is There a Season for Learning?" *Child Development* 58:1151–60.

Higham, John. 1988. *Strangers in the Land: Patterns of American Nativism, 1860–1925*. 2d ed. New Brunswick: Rutgers University Press.

Hirschl, Thomas A., and Mark R. Rank. 1991. "The Effect of Population Density on Welfare Participation." *Social Forces* 70 (September): 225–35.

Hochschild, Jennifer L. 1981. *What's Fair?: American Beliefs about Distributive Justice*. Cambridge: Harvard University Press.

————. 1995. *Facing Up to the American Dream: Race, Class, and the Soul of the Nation*. Princeton: Princeton University Press.

Hoffer, Thomas, Andrew Greeley, and James S. Coleman. 1985. "Achievement Growth in Public and Catholic Schools." *Sociology of Education* 58 (1985): 74–97.

Hogan, Denis P., and Evelyn M. Kitagawa. 1985. "The Impact of Social Status, Family Structure, and Neighborhood on the Fertility of Black Adolescents." *American Journal of Sociology* 90: 825–55.

Holtzblatt, Janet, Janet McCubbin and Robert Gillette. 1994. "Promoting Work through the EITC." *National Tax Journal* 47 (September): 591–607.

Hotz, Joseph V., Lixin Xu, Marta Tienda, and Avner Ahituv. 1995. "The Returns to Early Work Experience in the Transition from School to Work." Paper presented at the Working Group on the Problems of Low-Income Populations, Institute for Research on Poverty, University of Wisconsin, Madison, February 1995.

Hout, Michael. 1988. "Expanding Universalism, Less Structural Mobility: The American Occupational Structure in the 1980s." *American Journal of Sociology* 93 (May): 1358–1400.

————. 1995. "The Politics of Mobility," pp. 301–25 in Alan C. Kerckhoff (ed.), *Generations and the Lifecourse*. Boulder: Westview Press.

Hout, Michael, Adrian E. Raftery, and Eleanor O. Bell. 1993. "Making the Grade: Educational Stratification in the United States, 1925–1989," pp. 25–50 in Yossi Shavit and Hans Peter Blossfeld (eds.), *Persistent Inequality: Changing Educational Attainment in 13 Countries*. Boulder: Westview Press.

Howell, David R. 1994. "The Skills Myth." *The American Prospect* 18 (Summer): 81–90.

Hoynes, Hilary W. 1994. "Does Welfare Play Any Role in Women's Family Decisions?" Discussion paper, Department of Economics, University of California, Berkeley.

Humphreys, Lloyd G. 1989. "Intelligence: Three Kinds of Instability and Their Consequences for Policy," pp. 193–216 in Robert L. Linn (ed.), *Intelligence: Measurement, Theory, and Public Policy*. Chicago: University of Chicago Press.

Hurst, Lindsay C., and David J. Mulhall. 1988. "Another Calendar Savant." *British Journal of Pscyhiatry* 152:274–77.

Huston, F. Allan. 1993. *Testing Testing: The Social Consequences of the Examined Life*. Berkeley: University of California Press.

Huston, James L. 1993. "The American Revolutionaries, the Political Economy of Aristocracy, and the American Concept of the Distribution of Wealth, 1765–1900." *American Historical Review* 98 (October): 1079–1105.

Husttman, Elizabeth, and Sylvia Fava (eds.) 1985. *Housing Needs and Policy Approaches: Trends in Thirteen Countries*. Durham: Duke University Press.

Irvine, S. H., and J. W. Berry. 1988. "The Abilities of Mankind: A Reevaluation," pp. 3–59 in S. H. Irvine and J. W. Berry (eds.), *Human Abilities in Cultural Context*. New York: Cambridge University Press.

Jackson, Kenneth. 1980. "Race, Ethnicity, and Real Estate Appraisal: The Home

Owners' Loan Corporation and the Federal Housing Administration," *Journal of Urban History* 6:419–52.

Jackson, Kenneth. 1981. "The Spacial Dimensions of Social Control: Race, Ethnicity, and Government Housing Policy in the United States," pp. 79–118 in Bruce M. Stave (ed.), *Modern Industrial Cities: History, Policy, and Survival*. Beverly Hills: Sage Publishers.

———. 1985. *The Crabgrass Frontier: The Suburbanization of the United States*. New York: Oxford University Press.

Jacoby, Russell, and Naomi Glauberman (eds.). 1995. *The Bell Curve Debate: History, Documents, Opinions*. New York: Times Books.

Jankowski, Martín Sánchez. 1990. *Islands in the Street: Gangs and American Urban Society*. Berkeley: University of California Press.

———. 1995. "The Rising Significance of Status in U.S. Race Relations," pp. 77–98 in Michael Peter Smith and Joe R. Feagin (eds.), *The Bubbling Cauldron: Race, Ethnicity, and the Urban Crisis*. Minneapolis: University of Minnesota Press.

Jargowsky, Paul A. 1994. "Ghetto Poverty among Blacks in the 1980s." *Journal of Policy Analysis and Management* 13, 2:283–310.

Jencks, Christopher, with Marshall Smith, Henry Ackland, Mary Jo Bane, David Cohen, Herbert Gintis, Barbara Heyns, and Stephan Michelson. 1972. *Inequality: A Reassessment of the Effect of Family and Schooling in America*. New York: Basic Books.

Jencks, Christopher. 1991. "Is the American Underclass Growing?" pp. 28–100 in Christopher Jencks and Paul E. Peterson (eds.), *The Urban Underclass*. Washington, D.C.: Brookings Institution.

Jencks, Christopher, and Paul E. Peterson (eds.). 1991. *The Urban Underclass*. Washington, D.C.: Brookings Institution.

Jensen, Arthur R. 1969. "How Much Can We Boost IQ and Scholastic Achievement?" *Harvard Educational Review* 39:1–123.

———. 1981. *Straight Talk about Mental Tests*. New York: The Free Press.

Johnson, Paul. 1978. *A Shopkeeper's Millenium: Society and Revivals in Rochester, New York, 1815–1837*. New York: Hill and Wang.

Jordan, Winthrop. 1977. *White over Black: American Attitudes toward the Negro, 1550–1812*. New York: Norton.

Kahn, Lawrence M. 1992. "Discrimination in Baseball," pp. 163–88 in Paul M. Sommers, *Diamonds are Forever: The Business of Baseball*. Washington, D.C.: Brookings Institution.

Kamin, Leon J. 1995. "Lies, Damned Lies, and Statistics," pp. 81–105 in Russell Jacoby and Naomi Glauberman (eds.), *The Bell Curve Debate*. New York: Times Books.

———. 1995. "The Pioneers of IQ Testing," pp. 476–509 in Russell Jacoby and Naomi Glauberman (eds.), *The Bell Curve Debate*. New York: Times Books.

Kardiner, Abram, and Lionel Ovesey. 1951. *The Mark of Oppression: A Psycho-social Study of the American Negro.* New York: Norton.

Karoly, Lynn A. 1993. "The Trend in Inequality among Families, Individuals, and Workers in the United States: A Twenty-Five Year Perspective," pp. 99–164 in Sheldon Danziger and Peter Gottschalk (eds.), *Uneven Tides: Rising Inequality in America.* New York: Russell Sage Foundation.

Karoly, Lynn A., and Gary Burtless. 1995. "Demographic Change, Rising Earnings Inequality, and the Distribution of Personal Well-Being, 1959–1989." *Demography* 32 (August): 379–405.

Kasarda, John D. 1995. "Industrial Restructuring and the Changing Location of Jobs," pp. 215–68 in Reynolds Farley (ed.), *State of the Union: America in the 1990s*, vol. 1. New York: Russell Sage Foundation.

Katz, Lawrence F., Gary W. Loveman, and David Blanchflower. 1993. "A Comparison of Changes in the Structure of Wages in Four OECD Countries." Revised version of paper prepared for the National Bureau of Economic Research Comparative Labor Markets Project Conference on "Differences and Changes in Wage Structures," Cambridge, Mass., July 1992.

Katz, Michael B. 1986. *In the Shadow of the Poor House: A Social History of Welfare in America.* New York: Basic Books.

Kelley, Jonathan, and M. D. R. Evans. 1993. "The Legitimation of Inequality: Occupational Earnings in Nine Nations." *American Journal of Sociology* 99 (July): 75–125.

Kerckhoff, Alan C. 1986. "Effects of Ability Grouping in British Secondary Schools." *American Sociological Review* 51:842–58.

———. 1995. "Institutional Arrangements and Stratification Processes in Industrial Societies." *Annual Review of Sociology* 15:323–47.

Keyssar, Alexander. 1986. *Out of Work: The First Century of Unemployment in Massachusetts.* New York: Cambridge University Press.

King, Martin Luther, Jr. 1964. *Why We Can't Wait.* New York: Signet Books.

Kirschenman, Joleen, and Kathryn M. Neckerman. 1991. " 'We'd Love to Hire Them, But . . .': The Meaning of Race for Employers," pp. 203–34 in Christopher Jencks and Paul E. Peterson (eds.), *The Urban Underclass.* Washington, D.C.: Brookings Institution.

Klich, L. Z. 1988. "Aboriginal Cognition and Psychological Science," pp. 427–52 in S. H. Irvine and J. W. Berry (eds.), *Human Abilities in Cultural Context.* New York: Cambridge University Press.

Kluegel, James R., György Csepeli, Tamás Kolosi, Anatal Örkény, and Mária Neményi. 1995. "Accounting for Rich and Poor: Existential Justice in Comparative Perspective," pp. 179–209 in James R. Kluegel, David S. Mason, and Bernd Wegener (eds.), *Social Justice and Political Change: Public Opinion in Capitalist and Post-Communist States.* New York: Aldine de Gruyter.

Kluegel, James R., and Masaru Miyano. 1995. "Justice Beliefs and Support for the Welfare State in Advanced Capitalism," pp. 81–107 in James R. Kluegel,

289

David S. Mason, and Bernd Wegener (eds.), *Social Justice and Political Change: Public Opinion in Capitalist and Post-Communist States*. New York: Aldine de Gruyter.

Kluegel, James R., and E. R. Smith. 1981. "Beliefs about Stratification." *Annual Review of Sociology* 7:29–56.

Kohn, Melvin L. (1969) 1977. *Class and Conformity: A Study in Values*. 2d ed. Chicago: University of Chicago Press.

Kohn, Melvin, Atsushi Naoi, Carrie Schoenbach, Carmi Schooler, and Kazimierz M. Slomczynski. 1990. "Position in the Class Structure and Psychological Functioning in the United States, Japan, and Poland." *American Journal of Sociology* 95 (January): 964–1008.

Kohn, Melvin L., and Carmi Schooler. 1973. "Occupational Experience and Psychological Functioning." *American Sociological Review* 38 (February): 97–118.

———. 1978. "The Reciprocal Effects of the Substantive Complexity of Work and Intellectual Flexibility: A Longitudinal Assessment." *American Journal of Sociology* 84 (July): 24–52.

Korenman, Sanders, and Christopher Winship. 1995. "A Reanalysis of *The Bell Curve*." NBER Working Paper Series, no. 5230. Cambridge, Mass.: National Bureau of Economic Research.

Kozol, Jonathan. 1991. *Savage Inequalities: Children in America's Schools*. New York: Crown.

Kranzler, John H., and Arthur R. Jensen. 1991. "The Nature of Psychometric g: Unitary Process or a Number of Independent Processes?" *Intelligence* 15: 397–422.

Kristof, Nicholas D. 1995. "Japanese Outcasts Better Off than in the Past but Still Outcast." *New York Times*, November 30:1.

Krueger, Alan B. 1993. "How Computers Have Changed the Wage Structure: Evidence from Microdata, 1984–1989." *Quarterly Journal of Economics* 108 (February): 33–60.

Krugman, Paul. 1994. "Technology's Revenge," *Wilson Quarterly* (Autumn): 56–64.

Ku, Leighton, Freya L. Sonenstein, and Joseph H. Pleck. 1993. "Neighborhood, Family, and Work: Influences on the Premarital Behaviors of Adolescent Males." *Social Forces* 72 (December): 479–503.

Kugelmass, Sol, Amia Lieblich, and Dorit Bossik. 1974. "Patterns of Intellectual Ability in Jewish and Arab Children in Israel." *Journal of Cross-Cultural Psychology* 5 (June): 184–98.

Kuhn, Thomas S. 1962. *The Structure of Scientific Revolutions*. Chicago: University of Chicago Press.

Kulikoff, Allan. 1989. "The Transition to Capitalism in Rural America." *William and Mary Quarterly*, 3d ser., 46 (January): 120–44.

Lakatos, Imre. 1978. *The Methodology of Scientific Research Programmes*. Cambridge: Cambridge University Press.

Lane, Charles. 1995 (1994). "Tainted Sources," pp. 125–39 in Russell Jacoby and Naomi Glauberman (eds.), *The Bell Curve Debate*. New York: Times Books.

Lears, Jackson. 1994. *Fables of Abundance: A Cultural History of Advertising in America*. New York: Basic Books.

Lebergott, Stanley. 1976. *The American Economy: Income, Wealth, and Want*. Princeton: Princeton University Press.

Lee, Changsoo, and George DeVos (eds.). 1981. *Koreans in Japan: Ethnic Conflict and Accommodation*. Berkeley: University of California Press.

Lee, Yongsook. 1991. "Koreans in Japan and the United States," pp. 139–65 in Margaret A. Gibson and John U. Ogbu (eds.), *Minority Status and Schooling: A Comparative Study of Immigrants and Involuntary Minorities*. New York: Garland.

Lemieux, Thomas. 1993. "Unions and Wage Inequality in Canada and the United States," pp. 69–108 in David Card and Richard Freeman (eds.), *Small Differences that Matter: Labor Market and Income Maintenance in Canada and the United States*. Chicago: University of Chicago Press.

Leonard, Jonathan S. 1987. "The Interaction of Residential Segregation and Employment Discrimination." *Journal of Urban Economics* 21:323–46.

Lerner, Preston. 1994. "Making Work Pay." *Washington Monthly* 26 (April): 27ff.

Levy, Frank. 1987. *Dollars and Dreams: The Changing American Income Distribution*. New York: Russell Sage Foundation.

———. 1995. "Incomes and Income Inequality," pp. 1–57 in Reynolds Farley (ed.), *State of the Union: America in the 1990s*. New York: Russell Sage Foundation.

Levy, Frank S., and Richard C. Michel. 1991. *The Economic Future of American Families: Income and Wealth Trends*. Washington, D.C.: Urban Institute Press.

Levy, Frank, and Richard J. Murnane. 1992. "U.S. Earnings Levels and Earnings Inequality: A Review of Recent Trends and Proposed Explanations." *Journal of Economic Literature* 30 (September): 1333–81.

Lichter, Daniel T., and David Eggebeen. 1993. "Rich Kids, Poor Kids: Changing Income Inequality among American Children." *Social Forces* 71 (March): 761–80.

Lichter, Daniel T., Felecia B. LeClere, Diane K. McLaughlin. 1991. "Local Marriage Markets and the Marital Behavior of Black and White Women." *American Journal of Sociology* 96:843–67.

Lichter, Daniel T., Diane K. McLaughlin, George Kephart, and David J. Landry. 1992. "Race and the Retreat from Marriage: A Shortage of Marriageable Men?" *American Sociological Review* 57 (December): 781–99.

Lieberson, Stanley. 1963. *Ethnic Patterns in American Cities*. New York: Free Press of Glencoe.

———. 1973. "Generational Differences among Blacks in the North." *American Journal of Sociology* 79:550–65.

———. 1980. *A Piece of the Pie: Blacks and White Immigrants since 1880*. Berkeley: University of California Press.

Lieblich, Amia, et al. 1975. "Patterns of Intellectual Ability in Jewish and Arab Children in Israel: II. Urban Matched Samples," *Journal of Cross-cultural Psychology* 6 (June): abstract.

Lin, Nan, and Mary Dumin. 1986. "Access to Occupations Through Social Ties." *Social Networks* 8 (December): 365–85.

Lindenberger, Ulman, and Paul B. Baltes. 1994. "Sensory Functioning and Intelligence in Old Age: A Strong Connection." *Psychology and Aging* 9 (September): 339–55.

Lindert, Peter H. 1994. "The Rise of Social Spending." *Explorations in Economic History* 31:1–37.

Litwack, Leon F. 1979. *Been in the Storm So Long: The Aftermath of Slavery.* New York: Knopf.

Lockridge, Kenneth A. 1970. *A New England Town: The First Hundred Years, Dedham, Massachusetts, 1636–1736.* New York: Norton.

Logan, John R., and Richard D. Alba. 1993. "Locational Returns to Human Capital: Access to Suburban Community Resources." *Demography* 30 (May): 243–68.

Lucas, Samuel R. 1994. "Effects of Race and Gender Discrimination in the United States, 1940–1980." Ph.D. dissertation, University of Wisconsin.

———. 1995. "Educational Transitions of 1980 Sophomores: Background, Achievement, and Delinquency." Paper presented at the annual meeting of the American Sociological Association, Washington, D.C., August.

Lucas, Samuel R., and Adam Gamoran. 1991. "Race and Track Assignment: A Reconsideration with Course-Based Indicators of Track Locations." Paper presented at the annual meeting of the American Sociological Association, Cincinnati, August.

Lynn, Richard, Susan Hampton, and Mary Magee. 1984. "Home Background, Intelligence, Personality and Education as Predictors of Unemployment in Young People." *Personality and Individual Differences* 5:549–57.

McClosky, Herbert, and John Zaller. 1984. *The American Ethos: Public Attitudes Toward Capitalism and Democracy.* Cambridge: Harvard University Press.

McFate, Katherine, Timothy Smeeding, and Lee Rainwater. 1995. "Markets and States: Poverty Trends and Transfer System Effectiveness in the 1980s," pp. 29–66 in Katherine McFate, Roger Lawson, and William Julius Wilson (eds.), *Poverty, Inequality, and the Future of Social Policy.* New York: Russell Sage Foundation.

McLanahan, Sara, and Gary Sandefur. 1994. *Growing up with a Single Parent: What Hurts, What Helps.* Cambridge: Harvard University Press.

McWilliams, Carey. 1971 (1939). *Factories in the Field: The Story of Migratory Farm Labor in California.* Santa Barbara: Peregrine Publishers.

Mare, Robert D., 1991. "Five Decades of Educational Assortative Marriage." *American Sociological Review* 56 (February): 15–32.

———. 1995. "Changes in Educational Attainment and School Enrollment," pp. 155–213 in Reynolds Farley (ed.), *State of the Nation: America in the 1990s,* vol. 1. New York: Russell Sage Foundation.

Marshall, Jonathan. 1995. "Wage Picture May Be Brighter." *San Francisco Chronicle*, November 6: E1.

Massey, Douglas S., Gretchen A. Condran, and Nancy A. Denton. 1987. "The Effect of Residential Segregation on Social and Economic Well-Being." *Social Forces* 66 (September): 29–56.

Massey, Douglas S., and Nancy Denton. 1993. *American Apartheid: Segregation and the Making of the Underclass.* Cambridge: Harvard University Press.

Massey, Douglas S., and Mitchell L. Eggers. 1990. "The Ecology of Inequality: Minorities and the Concentration of Poverty, 1970–1980." *American Journal of Sociology* 96 (March): 1153–88.

Massey, Douglas S., and Eric Fong. 1990. "Segregation and Neighborhood Quality: Blacks, Hispanics, and Asians in the San Francisco Metropolitan Area." *Social Forces* 69 (September): 15–32.

Massey, Douglas S., and Andrew B. Gross. 1991. "Explaining Trends in Racial Segregation." *Urban Affairs Quarterly* 27 (September): 13–35.

Massey, Douglas S., Andrew B. Gross, and Kumiko Shibuya. 1994. "Migration, Segregation, and the Geographic Concentration of Poverty." *American Sociological Review* 59 (June): 425–45.

Mayer, Susan. 1995. "A Comparison of Poverty and Living Conditions in the United States, Canada, Sweden, and Germany," pp. 109–52 in Katherine McFate, Roger Lawson, and William Julius Wilson (eds.), *Poverty, Inequality, and the Future of Social Policy.* New York: Russell Sage Foundation.

Mayer, Susan, and Christopher Jencks. 1995. "War on Poverty: No Apologies, Please." *New York Times*, November 9.

Mickelson, Roslyn Arlin. 1990. "The Attitude-Achievement Paradox among Black Adolescents." *Sociology of Education* 63:44–61.

Miller, Leon K. 1987. "Developmentally Delayed Musical Savant's Sensitivity to Tonal Structure." *American Journal of Mental Deficiency* 91:467–71.

Milofsky, Carl. 1989. *Testers and Testing: The Sociology of School Psychology.* New Brunswick, N.J.: Rutgers University Press.

Mishel, Lawrence, and David M. Frankel. 1991. "Hard Times for Working America," *Dissent* (Spring): 282–85.

Moffitt, Robert. 1992. "Incentive Effects of the U.S. Welfare System." *Journal of Economic Literature* 30:1–61.

Monkkonen, Eric (ed.). 1984. *Walking to Work: Tramps in America, 1790–1935.* Lincoln: University of Nebraska Press.

Montejano, David. 1987. *Anglos and Mexicans in the Making of Texas, 1836–1986.* Austin: University of Texas Press.

Moore, Steven, and Dean Stansel. 1995. "Ending Corporate Welfare as We Know It," Cato Institute, draft report, March 6.

Mosle, Sara. 1995. "Dissed." *The New Yorker.* September 11: 8–9.

Müller, Walter, and Wolfgang Karle. 1993. "Social Selection in Educational Systems in Europe." *European Sociological Review* 9:1–23.

Murnane, Richard J., John B. Willett, and Frank Levy. 1995. "The Growing Impor-

293

tance of Cognitive Skills in Wage Determination." *Review of Economics and Statistics* 77 (May): 251–66.

Murray, Charles. 1995. "*The Bell Curve* and Its Critics." *Commentary* (May): 23–30.

Murray, Charles, and Richard Herrnstein. 1995. "Race, Genes, and I.Q.—An Apologia." *The New Republic*, October 31: 27–37.

Myers, Dowell, and Jennifer R. Wolch. 1995. "The Polarization of Housing Status," pp. 269–334 in Reynolds Farley (ed.), *State of the Union: America in the 1990s*, vol. 1. New York: Russell Sage Foundation.

Naoi, Michiko, and Carmi Schooler. 1990. "Psychological Consequences of Occupational Conditions among Japanese Wives." *Social Psychology Quarterly* 53 (June): 100–16.

Neal, Derek A., and William R. Johnson. 1994. "The Role of Pre-Market Factors in Black-White Wage Differences." Seminar on Meritocracy and Equality, University of Chicago.

Newman, Katherine. 1993. *Declining Fortunes: The Withering of the American Dream*. New York: Basic Books.

Newman, Katherine, and Chauncy Lennon. 1995. "The Job Ghetto." *The American Prospect* 22 (Summer): 66–67.

Nisbett, Richard. 1995. "Race, IQ, and Scientism," pp. 36–57 in Steven Fraser (ed.), *The Bell Curve Wars*. New York: Basic Books.

Novack, Janet. 1994. "The Worm in the Apple." *Forbes*, November 7: 96ff.

Oakes, J. 1985. *Keeping Track: How Schools Structure Inequality*. New Haven: Yale University Press.

Ogbu, John U. 1978. *Minority Education and Caste: The American System in Cross-cultural Perspective*. New York: Academic Press.

———. 1994. "Minority Status, Cultural Frame of Reference, and Schooling," pp. 61–83 in D. Keller-Cohen (ed.), *Literacy: Inter-Disciplinary Perspectives*. Cresskill, N.J.: Hampton.

———. 1994. "Racial Stratification and Education in the United States: Why Inequality Persists." *Teachers College Record* 96 (Winter): 264–98.

———. 1995. "Community Forces and Minority Educational Strategies." Final Report no. 1, Department of Anthropology, University of California, Berkeley (March).

O'Higgins, Michael, Günther Schmaus, and Geoffrey Stephenson. 1990. "Income Distribution and Redistribution," pp. 20–56 in Timothy M. Smeeding, Michael O'Higgins, and Lee Rainwater (eds.), *Poverty, Inequality, and Income Distribution in Comparative Perspective*. Washington, D.C.: Urban Institute Press, 1990.

Oliver, Melvin L., and Thomas M. Shapiro. 1995. *Black Wealth/White Wealth: A New Perspective on Racial Inequality*. New York: Routledge.

Orfield, Gary, and Carole Ashkinaze. 1991. *The Closing Door: Conservative Policy and Black Opportunity*. Chicago: University of Chicago Press.

Osberg, Lars. 1984. *Economic Inequality in the United States*. Armonk, N.Y.: M. E. Sharpe.

Padilla, Felix M. 1992. *The Gang as an American Enterprise*. New Brunswick: Rutgers University Press.

Parcel, Toby L., and Laura E. Geschwender. 1995. "Explaining Southern Disdavantage in Verbal Facility among Young Children." *Social Forces* 73 (March): 841–72.

Parcel, Toby L., and Elizabeth G. Menaghan. 1994. *Parents' Jobs and Children's Lives*. New York: Aldine de Gruyter.

Perkins, David. 1995. *Outsmarting IQ: The Emerging Science of Learnable Intelligence*. New York: Free Press.

Persson, Torsten, and Guido Tabellini. 1994. "Is Inequality Harmful for Growth?" *American Economic Review* 84 (June): 600–21.

Peterson, Jonathan. 1995. "GOP Seeking to Curb Tax Break for Poor." *Los Angeles Times*, May 13: A1.

Phillips, Kevin. 1990. *The Politics of Rich and Poor: Wealth and the American Electorate in the Reagan Aftermath*. New York: Random House.

———. 1994. *Boiling Point: Democrats, Republicans and the Decline of Middle-Class Prosperity*. New York: Harper Collins.

Porter, Michael. 1992. *Capital Choices: Changing the Way America Invests in Industry*. Washington, D.C.: Council on Competitiveness.

Quadagno, Jill. 1994. *The Color of Welfare: How Racism Undermined the War on Poverty*. New York: Oxford University Press.

Quane, James N., and Bruce H. Rankin. 1995. "Does Living in a Poor Neighborhood Affect 'Mainstream' Goals?" Manuscript, Center for the Study of Urban Inequality, University of Chicago.

Rader, Benjamin G. 1994. *Baseball: A History of America's Game*. Urbana: University of Illinois Press.

Rainwater, Lee. 1974. *What Money Buys: Inequality and the Social Meanings of Income*. New York: Basic Books.

Rainwater, Lee, and Timothy Smeeding. 1995. "Doing Poorly: The Real Income of American Children in Comparative Perspective." Working Paper no. 127, Luxembourg Income Study. Syracuse, N.Y.: Maxwell School of Citizenship and Public Affairs, August.

Rangari, Ashok. 1987. "Caste Affiliation, Sex, Area of Residence and Socioeconomic Status as Sources of Variation in Intelligence of Students." *Indian Psychological Review* 32, 5–6:43–49.

Raven, John. 1989. "The Raven Progressive Matrices: A Review of National Norming Studies and Ethnic and Socioeconomic Variation within the United States." *Journal of Educational Measurement* 26 (Spring): 1–16.

Restak, Richard M. 1994. *The Modular Brain: How New Discoveries in Neuroscience Are Answering Age-Old Questions about Memory, Free Will, Consciousness, and Personal Identity*. New York: Scribners'.

Roediger, David R. 1991. *The Wages of Whiteness: Race and the Making of the American Working Class.* New York: Verso, 1991.

Rogin, Michael Paul. 1975. *Fathers and Children: Andrew Jackson and the Subjugation of the American Indian.* New York: Knopf.

Rohlen, Thomas. 1981. "Education, Policies, and Prospects," pp. 182–222 in Changsoo Lee and George A. DeVos (eds.), *Koreans in Japan.* Berkeley: University of California Press.

Roos, Patricia A. 1990. "Hot Metal to Electronic Composition: Gender, Technology, and Social Change," pp. 275–98 in Barbara Reskin and Patricia Roos (eds.), *Job Queues, Gender Queues.* Philadelphia: Temple University Press.

Rosenbaum, Emily. 1994. "The Constraints on Minority Housing Choices, New York City 1978–1987." *Social Forces* 72 (March): 725–47.

Rosenbaum, James E. 1980. "Track Misperceptions and Frustrated College Plans: An Analysis of the Effects of Tracks and Track Perceptions in the National Longitudinal Survey." *Sociology of Education* 53:74–88.

Rosenbaum, James E., and Susan J. Popkin. 1990. "Why Don't Welfare Mothers Get Jobs? A Test of the Culture of Poverty and Spatial Mismatch Hypotheses." Center for Urban Affairs and Policy Research, Northwestern University.

Ruhul Amin and A. G. Mariam. 1987. "Racial Differences in Housing: An Analysis of Trends and Differentials, 1960–1978." *Urban Affairs Quarterly* 22 (March): 363–76.

Ryscavage, Paul. 1995. "A Surge in Growing Income Inequality?" *Monthly Labor Review* (August): 51–61.

St. George, Ross. 1971. "Cognitive Ability Assessment in New Zealand: Some Remarks," *New Guinea Psychologist* 3 (August 1971): 42–46.

Samuelson, Robert J. 1995. "Wages, Prices, and Profits." *Newsweek* September 25: 63.

Sandhu, T. S. 1986. "A Study of Caste Differences in Intelligence and Academic Achievement." *Journal of Psychological Researches* 30 (January): 30–33.

Sautter, Udo. 1991. *Three Cheers for the Unemployed: Government and Unemployment before the New Deal.* New York: Cambridge University Press.

Schaie, K. Warner. 1984. "Midlife Influences upon Intellectual Functioning in Old Age." *International Journal of Behavioral Development* 7:463–78.

———. 1990. "Intellectual Development in Adulthood," pp. 291–309 in James E. Birren and K. Warner Schaie (eds.), *The Handbook of the Psychology of Aging,* 3d ed. New York: Academic Press.

———. 1994. "The Course of Adult Intellectual Development." *American Psychologist* 49 (April): 304–13.

Schneider, Barbara, and Yongsook Lee. 1990. "A Model for Academic Success: The School and Home Environment of East Asian Students." *Anthropology and Education Quarterly* 21 (December): 358–77 (abstract).

Schneiderman, Anders. 1994. "The Hidden Handout: Housing and the Rise and Fall of the U.S. Welfare State." Ph.D. dissertation, University of California, Berkeley.

Schoenfeld, A. H. 1985. *Mathematical Problem-Solving*. New York: Academic Press.

Scholz, John Karl. 1994. "The Earned Income Tax Credit: Participation, Compliance, and Antipoverty Effectiveness." *National Tax Journal* 47 (March): 63–87.

Schooler, Carmi. 1989. "A Sociological Perspective on Intellectual Development." Presentation at the biennial meeting of the Society for Research in Child Development, Kansas City, April.

———. 1990. "Psychosocial Factors and Effective Cognitive Functioning in Adulthood," pp. 347–58 in James E. Birren and K. Warner Schaie (eds.), *The Handbook of the Psychology of Aging*, 3d ed. New York: Academic Press.

Schor, Juliet. 1992. *The Overworked American: The Unexpected Decline of Leisure*. New York: Basic Books.

Schorr, Lisbeth B., and Daniel Schorr. 1988. *Within Our Reach: Breaking the Cycle of Disadvantage*. New York: Anchor Press/Doubleday.

Schraagen, Jan Maarten. 1993. "How Experts Solve a Novel Problem in Experimental Design." *Cognitive Science* 17:285–309.

Schultz, Paul T. 1994. "Marital Status and Fertility: Welfare and Labor Market Effects." *Journal of Human Resources* 29:637–69.

Schuman, Howard, Charlotte Steeh, and Lawrence Bobo. 1985. *Racial Attitudes in America*. Cambridge: Harvard University Press.

Schwarz, John E. 1983. *America's Hidden Success: A Reassessment of Twenty Years of Public Policy*. New York: Norton.

Schwarz, John E., and Thomas J. Volgy. 1992. "Out of Line." *The New Republic* (22 November): 16–17.

Science for the People. 1974. "IQ" 6 (March).

Selfe, Lorna. 1977. *Nadia: A Case of Extraordinary Drawing Ability in an Autistic Child*. London: Academic Press.

Shamas, Carole. 1993. " A New Look at Long-term Trends in Wealth Inequality in the United States," *American Historical Review* 98 (April): 412–31.

Shapiro, Robert J. 1995. *Cut-and-Invest: A Budget Strategy for the New Economy*, Progressive Policy Institute, Policy Report no. 23, March.

Shimahara, A. 1991. "Social Mobility and Education: Burakumin in Japan," pp. 327–56 in Margaret A. Gibson and John U. Ogbu (eds.), *Minority Status and Schooling: A Comparative Study of Immigrants and Involuntary Minorities*. New York: Garland.

Shyam, Radhe. 1986. "Variations in Concentration of 'g' Level Abilities among Different Groups." *Journal of Personality & Clinical Studies* 2 (September): 123–26 (abstract).

Singer, Jerome L., and Dorothy G. Singer. 1984. "Psychologists Look at Television: Cognitive, Developmental, Personality, and Social Policy Implications," pp. 488–506 in Stella Chess and Alexander Thomas (eds.), *Annual Progress in Child Psychiatry and Child Development 1984*. New York: Brunner/Mazel.

Smeeding, Timothy M. 1992. "Why the U.S. Antipoverty System Doesn't Work Very Well," *Challenge* 35 (January): 30–36.

Smeeding, Timothy M., Michael O'Higgins, and Lee Rainwater (eds.). 1990. *Poverty, Inequality, and Income Distribution in Comparative Perspective.* Washington, D.C.: Urban Institute Press.

Smeeding, Timothy, Barbara Boyle Torrey, and Martin Rein. 1988. "Patterns of Income and Poverty: The Economic Status of Children and the Elderly in Eight Countries," pp. 89–120 in John L. Palmer, Timothy Smeeding, and Barbara Boyle Torrey (eds.), *The Vulnerable.* Washington, D.C.: Urban Institute Press.

Smith, James P., and Finis R. Welch. 1994 (1989). "Black Economic Progress after Myrdal," pp. 155–81 in Paul Burstein (ed.), *Equal Employment Opportunity: Labor Market Discrimination and Public Policy.* New York: Aldine de Gruyter.

Smith, Neil, and Iantha-Maria Tsimpli. 1995. *The Mind of a Savant: Language Learning and Modularity.* London: Basil Blackwell.

Smith, Tom W. 1990. "Social Inequality in Cross-National Perspective," pp. 21–29 in Duane F. Alwin et al., *Attitudes to Inequality and the Role of Government.* Rijswijk: Sociaal en Cultureel Planbureua; Alphen aan des Rijn: Samsom (Netherlands).

Smolensky, Eugene, Sheldon Danziger, and Peter Gottschalk. 1988. "The Declining Significance of Age in the United States: Trends in the Well-Being of Children and the Elderly Since 1939," pp. 29–54 in John L. Palmer, Timothy Smeeding, and Barbara Boyle Torrey (eds.), *The Vulnerable.* Washington, D.C.: Urban Institute Press.

Sniderman, Paul M., and Thomas Piazza. 1993. *The Scar of Race.* Cambridge, Mass.: Harvard University Press.

Soltow, Lee. 1993. "Wealth and Income Distribution," pp. 1517–32 in Mary Kupiec Cayton, Elliott J. Gorn, and Peter W. Williams (eds.), *Encyclopedia of American Social History.* New York: Charles Scribner's Sons.

Sowell, Thomas. 1977. "New Light on Black I.Q." *New York Times Magazine,* March 27: 56ff.

———. 1995. "Ethnicity and IQ." *The American Spectator* (February): 32–36.

Spearman, Charles. 1904 "General Intelligence, Objectively Determined and Measured." *American Journal of Psychology* 15:201–93.

Squires, Gregory D., and William Velez 1987. "Insurance Redlining in the Transformation of the Urban Metropolis." *Urban Affairs Quarterly* 23 (September): 63–83.

Stack, Carol B. 1974. *All Our Kin: Strategies for Survival in a Black Community.* New York: Harper & Row.

Steele, Claude. 1992. "Race and the Schooling of Black Americans." *Atlantic Monthy* 269 (April): 68ff.

———. 1993. "Protective Dis-Identification and Academic Performance," Research proposal to National Institute of Mental Health, Stanford University.

Steele, Claude, and Joshua Aronson. 1995. "Contending with a Stereotype: African-

American Intellectual Test Performance and Stereotype Vulnerability." Seminar on Meritocracy and Equality, University of Chicago, May.

Steinfels, Peter. 1995. "As Government Aid Evaporates, How Will Religious and Charity Organizations Hold Up?" *New York Times*, October 28: 7.

Sternberg, Robert J. 1988. *The Triarchic Mind: A New Theory of Human Intelligence*. New York: Viking.

Stevenson, Harold W., Chuansheng Chen, and Shing-Ying Lee. 1993. "Mathematics Achievement of Chinese, Japanese, and American Children: Ten Years Later." *Science* 259 (January 1): 53–58.

Storfer, Miles D. 1990. *Intelligence and Giftedness: The Contributions of Heredity and Early Environment*. San Francisco: Jossey-Bass.

Story, Ronald. 1993. "The Aristocracy of Inherited Wealth," pp. 1533–40 in Mary Kupiec Cayton, Elliott J. Gorn, and Peter W. Williams (eds.), *Encyclopedia of American Social History*. New York: Charles Scribner's Sons.

Sugrue, Thomas. 1993. "The Structures of Urban Poverty: The Reorganization of Space and Work in Three Periods of American History," pp. 85–117 in Michael B. Katz (ed.), *The "Underclass" Debate: Views from History*. Princeton: Princeton University Press.

Sullivan, Teresa A., Elizabeth Warren, and Jay L. Westbrook. 1989. *As We Forgive Our Debtors: Bankruptcy and Consumer Credit in America*. New York: Oxford University Press.

Swidler, Ann. 1992. "Inequality and American Culture." *American Behavioral Scientist* 35 (March/June): 606–29.

Taylor, Paul S. 1970. *Mexican Labor in the United States*. New York, Arno Press.

Thomas, Rich. 1995. "Rising Tide Lifts the Yachts: The Gap between Rich and Poor Has Widened, but There Are Some Comforting Twists," *Newsweek*, May 1: 62D.

Tienda, Marta, Katherine Donato, and Hector Cordero-Guzman. 1992. "Schooling, Color, and the Labor Force Activity of Women." *Social Forces* 71:365–79.

Tobey, Ronald, Charles Wetherell, and Jay Brigham. 1990. "Moving Out and Settling In: Residential Mobility, Home Owning, and the Public Enframing of Citizenship, 1921–1950." *American Historical Review* 95 (December): 1395–1423.

Trachtenberg, Alan. 1982. *The Incorporation of America: Culture and Society in the Gilded Age*. New York: Hill and Wang.

Trattner, Walter I. 1994. *From Poor Law to Welfare State: A History of Social Welfare in America*. 5th ed. New York: Free Press.

Triandis, Harry C. 1990. "Cross-Cultural Studies of Individualism and Collectivism," pp. 41–33 in J. Berman (ed.), *Nebraska Symposium on Motivation, 1989*. Lincoln: University of Nebraska Press.

Tucker, William H. 1994. *The Science and Politics of Racial Research*. Urbana: University of Illinois Press.

Uchitelle, Louis. 1995. "Union Goal of Equality Fails the Test of Time." *New York Times*, July 9: 1, 10.

U.S. Bureau of Labor Statistics. 1996. "New Productivity Data: 1949–1994." BLS home page: www.bls.gov/data.

U.S. Bureau of the Census. 1977. *Historical Statistics of the United States*. Washington, D.C.: USGPO.

U.S. Bureau of the Census. 1992. "Studies in the Distribution of Income." Current Population Reports P60–183. Washington, D.C.: USGPO.

———. 1992. "Workers with Low Earnings: 1963 to 1990." Current Population Reports, series P-60, no. 178. Washington, D.C.: USGPO.

———. 1993. "Measuring the Effect of Benefits and Taxes on Income and Poverty: 1992." Current Population Reports, series P-60–186RD. Washington, D.C.: USGPO.

———. 1994. "Debt on Single-Family Prpoerties." Press release of December 5. Census home page: http://gopher.census.gov: 70/0/Bureau/Pr/Subject/ House/Cb 94–182 txt.

———. 1994. "Health Insurance Coverage—1993." Statistical Brief SB/94–28, October.

———. 1994. "Poverty—Long and Short Term." Bureau of the Census Statistical Brief, December 1994.

———. 1994. *Statistical Abstract of the United States 1994*. Washington, D.C.: USGPO.

———. 1995. "Asset Ownership of Households: 1993." Census home page: www.census.gov/ftp/pub/hhes/wealth.November.

———. 1995. "Dynamics of Economic Well-Being: Labor Force, 1991 to 1993." Census home page: www.census.gov/ftp/pub/hhes//laborfor/dewb9193/highlight .hmtl.October.

———. 1995. "Housing Affordability." Census home page: www.census.gov/ftp/ hhes/www/hsgaffrd.html. October 5.

———. 1995. "Income and Job Mobility in the Early 1990's," Statistical Briefs, U.S. Department of Commerce, Economics and Statistics Administration, March.

———. 1995. "Median Net Worth of Nation's Households Dropped," Press release, 18 April, distributed on Bureau of Census gopher; based on T. J. Eller, "Household Wealth and Asset Ownership: 1991," Publication P70–34, Washington: U.S. Bureau of the Census.

———. 1995. *Population Profile of the United States: 1995*. Current Population Reports, series P23–189. Washington: USGPO.

———. 1995. *Statistical Abstract of the United States 1995*. Washington, D.C.: USGPO.

van Valet, William (ed.). 1990. *International Handbook of Housing Policies*. New York: Greenwood Press.

Van Woodward, C. 1966. *The Strange Career of Jim Crow*. 2d ed. New York: Oxford University Press.

Vandal, Gilles. 1992. "The Nineteenth-Century Municipal Responses to the Problem of Poverty." *Journal of American History* 19 (November): 30–59.

Verba, Sidney, and Gary R. Orren. 1985. *Equality in America: The View from the Top*. Cambridge: Harvard University Press.

Verster, J. M., and R. J. Prinsloo. 1988. "The Diminishing Test Performance Gap between English Speakers and Afrikaans Speakers in South Africa," pp. 534–60 in S. H. Irvine and J. W. Berry (eds.), *Human Abilities in Cultural Context*. Cambridge: Cambridge University Press.

Wacquant, Loïc J. D. 1994. "Life in the Zone: The Social Art of the Hustler in the Contemporary Ghetto." Working Paper No. 49. New York: Russell Sage Foundation.

Wacquant, Loïc J. D., and William J. Wilson. 1989. "Poverty, Joblessness, and the Social Transformation of the Inner City," pp. 70–102 in Phoebe H. Cottingham and David Ellwood (eds.), *Welfare Policy for the 1990s*, Cambridge: Harvard University Press.

Wagner, Richard K. 1994. "Context Counts: The Case of Cognitive-Ability Testing for Job Selection," pp. 133–51 in Robert J. Sternberg and Richard K. Wagner (eds.), *Mind in Context: Interactionist Perspectives on Human Intelligence*. New York: Cambridge University Press.

Waters, Mary C., and Karl Eschenbach. 1995. "Immigration and Ethnic and Racial Inequality in the United States." *Annual Review of Sociology* 21:419–46.

Weakliem, David, Julia McQuillan, and Tracy Schauer. 1995. "Toward Meritocracy? Changing Social-Class Differences in Intellectual Ability." *Sociology of Education* 68 (October): 271–86.

Wegener, Bernd. 1991. "Job Mobility and Social Ties: Social Resources, Prior Job, and Status Attainment." *American Sociological Review* 56 (February): 60–71.

Weicher, John C. 1995. "Changes in the Distribution of Wealth: Increasing Inequality." *Federal Reserve Bank of St. Louis Review* 77 (January/February): 5–23.

Weinstein, Rhona S. 1976. "Reading Group Membership in First Grade: Teacher Behaviors and Pupil Experience over Time." *Journal of Educational Psychology* 68:103–16.

Weiss, Richard. 1988. *The American Myth of Success: From Horatio Alger to Norman Vincent Peale*. Urbana: University of Illinois Press.

Whyte, William Foote. 1993 (1943). *Street Corner Society: The Social Structure of an Italian Slum*. 4th ed. Chicago: University of Chicago Press.

Will, George. 1995. "What's Behind Income Disparity." *San Francisco Chronicle*, April 24: A15.

Williams, David R., and Chiquita Collins. 1995. "US Socioeconomic and Racial Differences in Health: Patterns and Explanations." *Annual Review of Sociology* 21: 349–86.

Williams, Terry M. 1989. *The Cocaine Kids: The Inside Story of a Teenage Drug Ring*. Reading, Mass.: Addison-Wesley.

Williams, Terry M., and William Kornblum. 1985. *Growing Up Poor*. Lexington, Mass.: Lexington Books.

Williamson, Jeffrey G., and Peter H. Lindert. 1980. *American Inequality*. New York: Academic Press.

———. 1980. "Long-Term Trends in American Wealth Inequality," pp. 9–94 in James D. Smith (ed.), *Modelling the Distribution and Intergenerational Transmission of Wealth*. Chicago: University of Chicago Press.

Willis, Paul E. 1977. *Learning to Labour: How Working Class Kids Get Working Class Jobs*. Farnborough, England: Saxon House.

Wilson, William J. 1987. *The Truly Disadvantaged*. Chicago: University of Chicago Press.

———. 1991. "Social Theory and the Public Agenda: The Challenge of Studying Inner-City Social Dislocations." *American Sociological Review* 56 (February): 1–16.

———. (ed.). 1989. *The Ghetto Underclass: Social Science Perspectives. The Annals* 501 (January).

Wolf, Stephanie Grauman. 1976. *Urban Village: Population, Community, and Family Structure in Germantown, Pennsylvania, 1683–1800*. Princeton: Princeton University Press.

Wolfe, Barbara L. 1994. "Reform of Health Care for the Nonelderly Poor," pp. 253–88 in Sheldon H. Danziger, Gary D. Sandefur, and Daniel H. Weinberg (eds.), *Confronting Poverty: Prescriptions for Change*. Cambridge: Harvard University Press.

Wolff, Edward N. 1995. "How the Pie Is Sliced: America's Growing Concentration of Wealth." *The American Prospect* 22 (Summer): 58–64.

———. 1995. *Top Heavy: A Study of the Increasing Inequality of Wealth in America*. New York: Twentieth-Century Fund Press.

Woodruff-Pak, Diana S. 1989. "Aging and Intelligence: Changing Perspectives in the Twentieth Century." *Journal of Aging Studies* 3, 2:91–118.

Wooldridge, Adrian. 1995. "Bell Curve Liberals: How the Left Betrayed I.Q.," *New Republic* 212 (February 27): 22–25.

❖ *Index* ❖

About the Authors

All of the authors are in the Department of Sociology at the University of California, Berkeley. CLAUDE S. FISCHER's books include *The Urban Experience* and *To Dwell among Friends: Personal Networks in Town and City*; MICHAEL HOUT is the author of *Following in Father's Footsteps: Occupational Mobility in Ireland* and *Mobility Tables*; MARTÍN SÁNCHEZ JANKOWSKI is the author of *City Bound: Urban Life and Political Attitudes among Chicano Youth* and *Islands in the Street: Gangs and American Urban Society*; SAMUEL R. LUCAS is completing a book on the effects of race and sex discrimination since 1940; ANN SWIDLER's books include *Organization without Authority: Dilemmas of Social Control in Free Schools* and *Habits of the Heart: Individualism and Commitment in American Life*; and KIM VOSS is the author of *The Making of American Exceptionalism: The Knights of Labor and Class Formation in the Nineteenth Century*.